Current Drying Processes

Edited by Israel Pala-Rosas

Published in London, United Kingdom

Supporting open minds since 2005

Current Drying Processes
http://dx.doi.org/10.5772/intechopen.78916
Edited by Israel Pala-Rosas

Contributors
Le Anh Duc, Keum Dong Hyuk, Chang Peng, Saeed Moghaddam, Artem Artyukhov, Jozef Bocko, Jan Krmela, Nadiia Artyukhova, Ruslan Ostroha, Mykola Yukhymenko, Margarita Castillo Téllez, Beatriz Castillo Téllez, José Andrés Alanís Navarro, Juan Carlos Ovando Sierra, Gerardo A. Mejia Pérez, Italo Pedro Bello Moreira, Edgar Ruperto Macías Ganchozo, Xavier Enrique Anchundia Muentes, Celio Danilo Bravo Moreira, Manuel Eduardo Anchundia Muentes, Hebert Edison Vera Delgado, Carlos Eduardo Anchundia Betancourt, Trung Thanh Bui

First published in London, United Kingdom, 2020 by IntechOpen
IntechOpen is the global imprint of INTECHOPEN LIMITED, registered in England and Wales, registration number: 11086078, 7th floor, 10 Lower Thames Street, London, EC3R 6AF, United Kingdom
Printed in Croatia

British Library Cataloguing-in-Publication Data
A catalogue record for this book is available from the British Library

Additional hard and PDF copies can be obtained from orders@intechopen.com

Current Drying Processes
Edited by Israel Pala-Rosas
p. cm.
Print ISBN 978-1-83880-109-0
Online ISBN 978-1-83880-110-6
eBook (PDF) ISBN 978-1-78985-861-7

For EU product safety concerns:
IN TECH d.o.o., Prolaz Marije Krucifikse Kozulić 3, 51000 Rijeka, Croatia, info@intechopen.com or visit our website at intechopen.com.

Meet the editor

Israel Pala-Rosas is a biochemical engineer at the Instituto Tecnológico de Tehuacán (Tehuacán, México), with a Master in chemical engineering from the Benemérita Universidad Autónoma de Puebla (Puebla, México) and a PhD in sciences in chemical engineering from the School of Chemical Engineering and Extractive Industries of the Instituto Politécnico Nacional (México). His interest is the research and development of catalytic and biotechnological processes for the transformation of biomass-derived molecules to compounds of technological and industrial interest. With this aim, Israel Pala-Rosas focuses his work on the synthesis, characterization, and testing of catalysts, as well as the design and analysis of chemical and biochemical reactors. Areas related to the catalytic processes, such as chemical thermodynamics, unit operations and economics, are also under his scope.

Contents

Preface

The transformation of a raw material in to a commodity or specialty product implies a series of steps in which the feedstock is subjected to physical and chemical changes. The stages in which only physical changes occur are known as unit operations, and may involve heat, mass, or both transport processes, as well as the presence of gaseous, liquid, or solid phases.

Drying operations have been widely employed in the industry to eliminate water from the process materials. Additionally, it also refers to the elimination of organic solvents and liquids from solids. It allows for the pretreatment of feedstocks with different moisture content, such as wet solids, gels, suspensions, and solutions, for their physical or chemical transformation. Drying processing also enables the post-treatment of products for their final presentation and packaging. In some industries, such as alimentary and pharmaceutical, drying is used as a preservation method of products since the enzymatic and microbial activity that causes undesirable chemical changes in organic matter is reduced with the decrease of water content.

From a process approach, the development of drying equipment requires the understanding and mathematical interpretation of the heat and mass transport phenomena for their proper design, construction, and operation, as well as an energy analysis that allows for the improvement of the process.

Under this scenario, *Current Drying Processes* presents recent advances in the development of drying operations for organic and inorganic materials dealing with the theoretical and experimental aspects of the most widely used technologies today, namely solar, convective, fluidized, and ultrasonic drying. The chapters present the elementary principles of each drying technology, such as energy transfer, hydrodynamics, and drying kinetics, as well as the importance of simulation and statistical methods for the analysis and improvement of the processes.

Israel Pala-Rosas
ESIQIE,
Instituto Politécnico Nacional,
México

Section 1

Solar Drying

Chapter 1

Kinetics of Drying Medicinal Plants by Hybridization of Solar Technologies

Margarita Castillo Téllez, Beatriz Castillo Téllez, José Andrés Alanís Navarro, Juan Carlos Ovando Sierra and Gerardo A. Mejia Pérez

Abstract

Historically, medicinal plants have always had an important place in medicine. Medicinal plants processing represents a great challenge, due to their compounds sensitive to the environmental conditions that surround and degrade them. Mostly of these plants require to be dry to preserve its safety and medicinal properties; therefore, for proper drying, it is necessary to use sustainable devices that protect the desirable characteristics of plants from direct radiation. In this work, the kinetics of dehydration of three medicinal plants are presented in an indirect solar dryer. In addition, the experimental results were adjusted to nine mostly used models, to estimate the drying conditions required to achieve a desired final moisture content. Modified Page and Page were the models with better fit to experimental results. Furthermore, a computational simulation of temperature evolution and distribution inside the dryer is presented. These results agree with those obtained experimentally.

Keywords: medicinal plants, indirect solar drying, mathematical model, colorimetric study, computational analysis

1. Introduction

The knowledge of medicinal plants extends to any part of the world where man has traditionally needed them to cure his illnesses. Thus, a mixture of magic and religion, combined with necessity and chance, trial and error, the passage of different cultures has created a whole knowledge of plant remedies that has been the basis of modern medicine [1]. However, many studies have been performed to test if they are truly effective and, as a result of these studies, the therapeutic use of many medicinal plants as substitutes for pharmaceutical medicines has been applied successfully to cure or relieve diseases [2].

According to the WHO, herbal medicines include herbs, herbal material, preparations, and herbal products, which contain as active ingredients parts of plants or other plant materials, and their use is well established and widely recognized as safe and effective [3].

It has been shown in studies and reviews that medicinal plants have various properties that cure, for example, anticancer and antiviral activities [4], antidiabetic properties [5], anti-dengue activities [6], infertility problems [7], contusion and swelling, plants that improves learning and retrieval processes, and facilitates memory retention, to treat Alzheimer disease, hepatitis, significant anesthetic activity, antiarrhythmic action and even anti-obesity [8–11].

In this work, three medicinal plants were selected, which for many years have been considered important because of the amount of medicinal properties they possess and because the literature has corroborated their effectiveness in various treatments.

One of the most used plants for medical purposes in the world is *Annona* (*Annona muricata* L.) which is a comestible tropical fruit widely cultivated through the world [12]. The roots of these species are used for their antiparasitic and pesticidal properties. Intensive chemical investigations of the leaves and seeds of this plant have resulted in the isolation of a great number of bioactive compounds which were found to display interesting biological including antitumor, anticancer, antiparasitic and pesticidal properties [13, 14]. Literature revealed that *Annona* fruit extracts possess antioxidant properties and were able to inhibit enzymes relevant to type-2 diabetes and hypertension [15].

On the other hand, also, *Moringa oleifera*, native to India, is fully utilized for its high nutritive values. The leaves have many minerals and vitamins and are utilized to treat malnutrition, to increase breast milk production in women, and as a powerful antioxidant, anticancer, anti-inflammatory, antidiabetic, and antimicrobial agent. There is literature that provides relaxing inducing properties [16–19].

Finally, Cymbopogon. It is native to Asia and grows in tropical and subtropical regions [19]. There are many medicinal uses of *Cymbopogon* because it has many functional properties. The leaves and the essential oil are consumed to help with dyspeptic disorders, colds, nervous conditions, and exhaustion. It was found that they have a potential anticarcinogenic action [20] and antimicrobial, sedative, spasmolytic, and carminative effects. It has also been shown to have antifungal, anti-inflammatory, antimutagenic, antimalarial, antinociceptive, antibacterial, and cholesterol reduction effects [21–26].

The preparation of medicinal plants depending on its use, is very important to reach the maximum potential action in the human body, but the most frequently performed preparation is drying the leaves and using them in teas or distilling them to produce essential oils.

Drying is fundamentally defined the elimination of the majority of plant moisture content, permitting the prevention of enzymatic and microbial activity and therefore preserving the product for extended shelf life. Moreover, the properties of medicinal plants are determined by their moisture content. Mostly of its preparation needs the elimination of water. Consequently, adequate dryers are very important, including control of temperature, velocity, and humidity values, and in many cases, direct solar radiation is forbidden to provide a rapid reduction in the moisture content without affecting the quality of medicinal plants. Drying process aids the marketing of plants, because drying allows to improve the distribution of the plants since it reduces both the weight and the volume, reducing the transport and storage needs [27].

Therefore, in this work, an indirect solar dehydrator was designed, with which the plant is protected from its high sensitivity to high temperatures and solar radiation.

2. Experimental study

2.1 Construction of the solar dryer

In this study, an indirect solar dryer type tunnel with solar technology hybridization was constructed (**Figure 1**): This dryer is hermetically sealed and is composed of three front sections. Each section contains two trays with a metallic

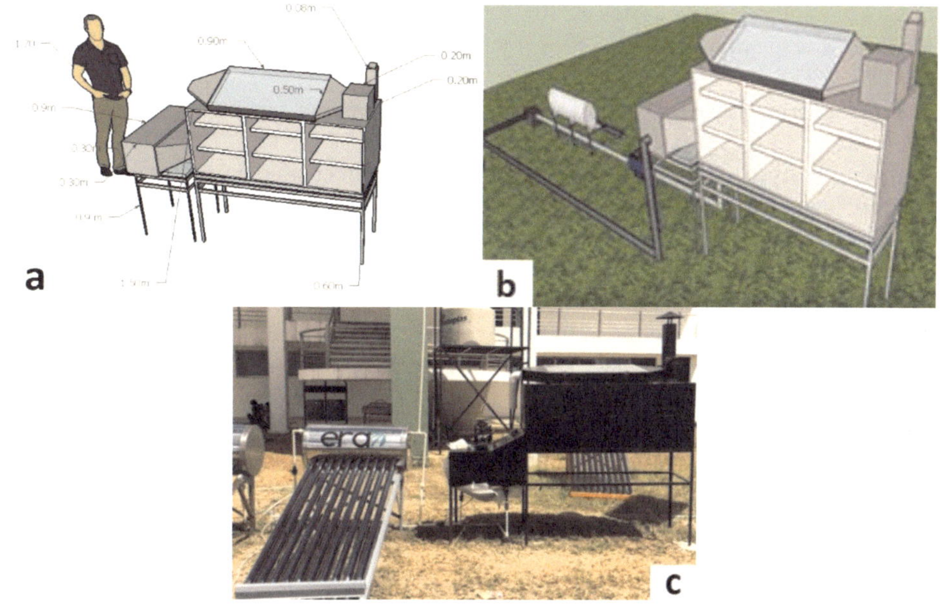

Figure 1.
Tunnel-type solar dryer: (a) main measurements and location of trays, (b) location of solar heater and solar air collector, (c) the indirect solar dryer in operation. Source: Own elaboration.

mesh to support the hot air flow evenly. Each tray has a separation of 20 cm, as shown in **Figure 1a**. The drying chamber is connected to a solar water heater by a hopper, which drives preheated air inside, supported by a heat exchanger. Additionally, the dryer has a solar air collector installed at the top. This collector contains a fan that extracts the ambient air through the interior and directs it, with an increase in temperature, to the interior of the drying chamber (**Figure 1b**). The solar dryer has a 1/2 hp. motor that extracts ambient air into the solar air collector and an electric pump that recirculates the water in the solar heater, both systems work with a 280 W solar panel. **Figure 1c** shows the indirect solar dryer in operation.

2.2 Measuring instruments

In order to ensure the reliability of the experimental results, different measuring equipment were used at the different stages of the experimentation; Boeco mark food moisture measurement scale, model BMA150, with an accuracy of ± 1 mg (0.001%) was used to determine the initial and final humidity of the leaves. The leaf sample of approximately 1.5 g was cut and placed in the analyzer. Water activity in the leaves under study was measured by a mark team Rotronic HygroPalm, with an accuracy of $\pm 0.01\%$ mg. The mean of three measurements was reported at a room temperature of $24.5 \pm 1°C$. The temperature and humidity inside the drying chambers were measured using a Brannan thermo-hygrometer with temperature and relative humidity accuracy of $\pm 1°C$ and $\pm$ 3%, respectively. The weight of the samples was measured using a Boeco weighing scale, model BPS40plus, which have an accuracy of ± 0.001 g.

The weather parameters were monitored by the weather station located in the Faculty of Engineering of the Autonomous University of Campeche, which consists of a LI-COR Pyranometer, with which the global solar irradiance was measured; the accuracy is Azimuth: $< \pm$ 1% on 360° to 45° of elevation. The relative humidity, ambient temperature, and wind velocity and direction were measured by NG Systems mark equipment, model RH-5X (accuracy $\pm 3\%$),

110S (accuracy ±1.1°C), Series #200P (accuracy ±3°), and P2546C-OPR (accuracy ±0.3 m/s), respectively.

2.3 Drying process and experimental setup

In this study, the kinetics of drying medicinal plants using an indirect solar dryer were analyzed experimentally to determine optimal operating conditions, evaluating the possibility of the integration of solar technologies in solar food drying. Samples of fresh leaves were cut from plantings on agricultural land, in the month of March 2019, in the city of Campeche, Campeche, Mexico. The experimental study was carried out from March 4 to April 26. The branches were cut, and the leaves were separated and selected to obtain a homogeneous group, based on maturity, color, freshness, and size. They were washed and weighed, and the width, length, and thickness were measured.

To analyze the initial and final humidity in the plants, the leaf sample of approximately 1.5 g was cut and placed in the analyzer. Water activity (a_w) is a parameter that determines the stability of the food with respect to the ambient humidity. It was measured for both fresh and dried leaves before and after the drying process using portable water activity meter.

The plants that were selected for experimentation were *Moringa oleifera*, *Annona muricata L.*, and *Cymbopogon citratus. Moringa oleifera*, is a plant that has generated great interest in recent years because of its attributed medicinal properties [28], contains more than 90 nutrients, different antioxidants, and all the 8 essential amino acids [29]. The main medicinal property of *Cymbopogon Citratus* is to be anti-inflammatory and antioxidant due to polyphenols, it contains, so it is used in cases of cancer and to combat arthritis among other properties [30].

Fresh and dried leaves were placed on a flat surface, and the colorimeter was placed on the leaves to measure color values. For each sample, the measurements were replicated four times, obtaining in each measurement the values L * (luminance), a * (brownness), and b * (darkness) [31].

The total color change (ΔE) was the parameter considered for the overall color difference evaluation, between a dried sample and the fresh leaf [32]:

$$\Delta E = \sqrt{(L_o - L)^2 + (a_o - a)^2 + (b_o - b)^2} \tag{1}$$

2.4 Mathematical modeling

In order to observe the drying behavior of medicinal plants, the moisture ratio against the drying time was plotted. The dry-based moisture ratio during the drying process was calculated using the following equation [33]:

$$MR_{db} = \frac{M_t}{M_0} \tag{2}$$

where M_t and M_0 are the moisture content in time t and the initial moisture content (kg water/kg dry matter), respectively.

2.4.1 Adjusting experimental drying curves

The experimental moisture-versus-time ratio was adjusted by nonlinear regression to nine thin-layer drying models (**Table 1**) for each drying kinetic of the

Model name	Model equation	Reference
Newton	$MR = \exp(-kt)$	[35]
Page	$MR = \exp(-kt^n)$	[36]
Modified page	$MR = \exp(-(kt)^n)$	[37]
Henderson and Pabis	$MR = a\ \exp(-kt)$	[38]
Logarithmic	$MR = a\exp(-kt) + c$	[39]
Two-term	$MR = a\exp(-k_0t) + b\ \exp(-k_1t)$	[40]
Two-term exponential	$MR = a\exp(-kt) + (1 - a)$	[41]
Wang and Singh	$MR = 1 + \mathrm{at} + \mathrm{b}t^2$	[42]
Weibull	$MR = \exp\lfloor-(t/b)^{\alpha}\rfloor$	[43]

Source: Own elaboration.

Table 1.
Mathematical models used to predict the drying kinetics of medicinal plants.

medicinal plants. The criteria used to select the model that best fit the data experimental were the coefficient of determination (R^2) and the root of the mean square error (RMSE), considering that values of R^2 superior to 0.95 RMSE values below 0.06 indicate good fit [34]. In addition, a lower value of X^2 is considered as indicative of better fit. The adjustment was solved for the calculation of the different parameters involved in the selected fit models using the DataFix software version 9.1.

3. Results and discussion

Moisture tests were performed on fresh and dehydrated leaves; in all cases, the average initial humidity was 73.8%, the maximum readings were 79.58%, and the minimum readings were 68.2%. These readings agree with those reported in the literature (Banchero, Carballo, & Telesca, 2007). The initial average a_w was 0.976, the minimum measured was 0.96, while the maximum was 0.98. The average final a_w in all cases was 0.44, the minimum measured reading was 0.33, and the maximum was 0.58. The final a_w values indicate that there is no possibility of microbial growth in the dehydrated product obtained [44].

3.1 Weather conditions

Figure 2 shows the change in the weather parameters during the test period with three sunny days. As can be seen, a maximum solar global irradiance of 952 W/m^2 was achieved, with the average maximum values ranging between 874 and 962 W/m^2. The average ambient temperature varied of 30°C y 35.7°C. On the other hand, the minimum RH (relative humidity) ranged between 44 and 46%; the maximum average on the test days ranged between 60 and 81%.

Figure 3 shows the temperatures inside the drying chamber, solar radiation, and the ambient temperature during the hours with the highest solar incidence on the selected day as an example. The drying chamber consists of three sections with similar temperatures. The highest temperature reaches 50.7°C.

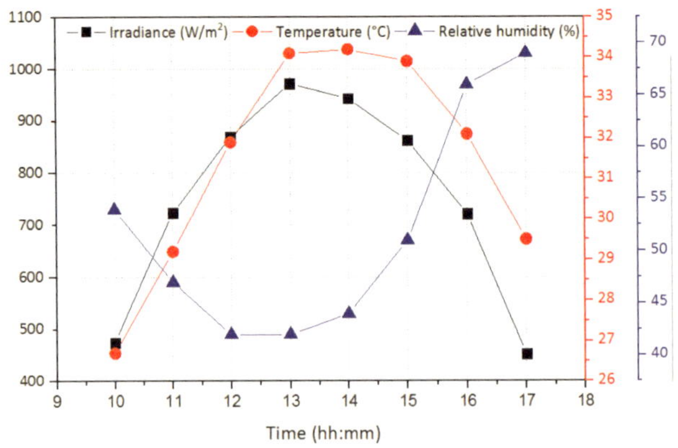

Figure 2.
Average solar irradiance, ambient temperature, and relative humidity. Source: Own elaboration.

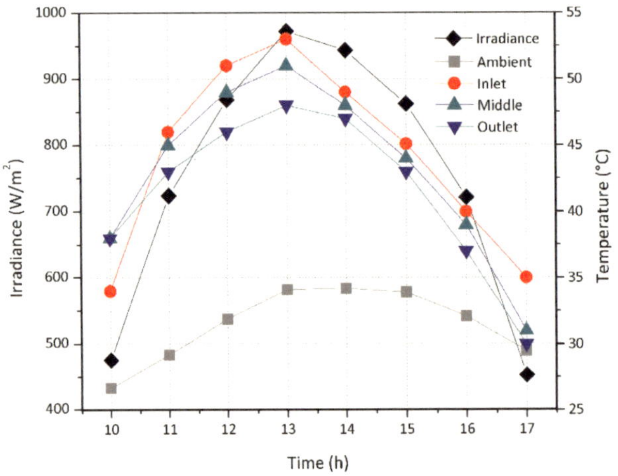

Figure 3.
Solar irradiance variation and temperature inside the drying chamber of the tunnel solar dryer. Source: Own elaboration.

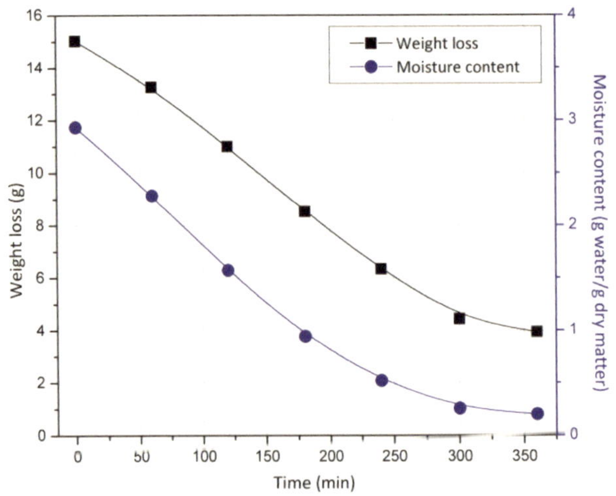

Figure 4.
Weight loss and moisture content depending on the drying time for Moringa oleifera *leaves. Source: Own elaboration.*

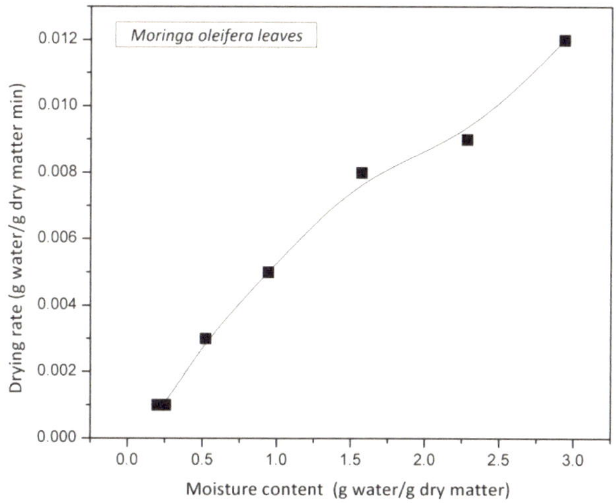

Figure 5.
Drying rate as function of moisture content for a Moringa oleifera *leaves. Source: Own elaboration.*

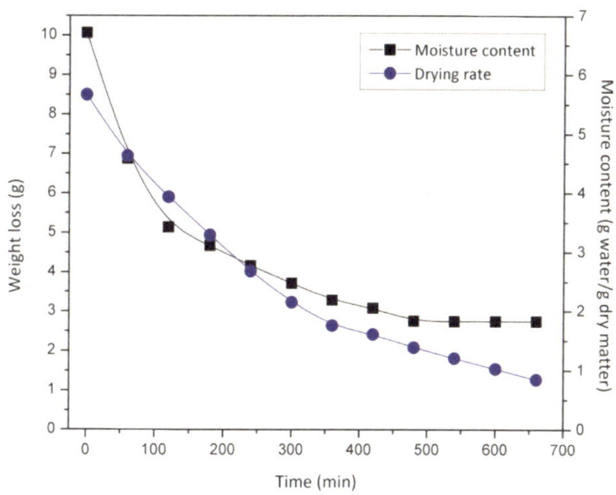

Figure 6.
Weight loss and moisture content depending on the drying time for Annona muricata *leaves. Source: Own elaboration.*

3.2 Solar drying kinetics

3.2.1 Moringa oleifera *leaves*

The weight loss stabilized upon reaching this sample 3.9 g; the final drying time was 360 min. It can be seen in **Figure 4** that the moisture content started at 2.93 g water/g dry matter, ending between 0.201 and 0.256. **Figure 5** shows the moisture content as a function of the drying rate; in this case no constant rate period was observed. The highest drying rate was found with 0.012 g water/g dry matter minute.

3.2.2 Annona muricata *leaves*

In the case of *Annona muricata* leaves, the total drying time was 500 min, (**Figure 6**); the initial moisture content was 5.667 g water/g dry matter, reaching a

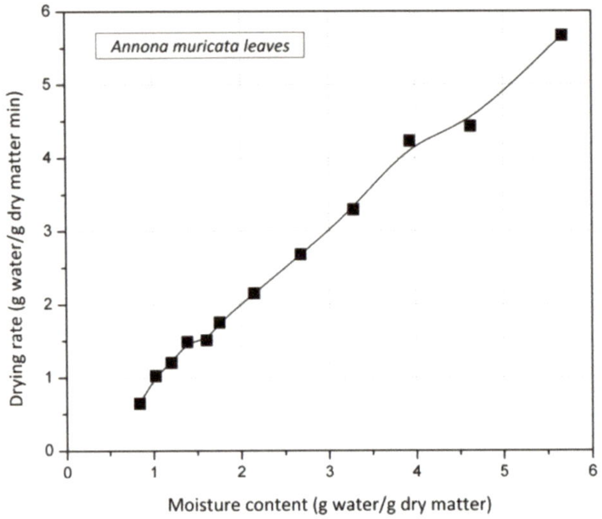

Figure 7.
Drying rate as function of moisture content for an Annona muricata *leaves. Source: Own elaboration.*

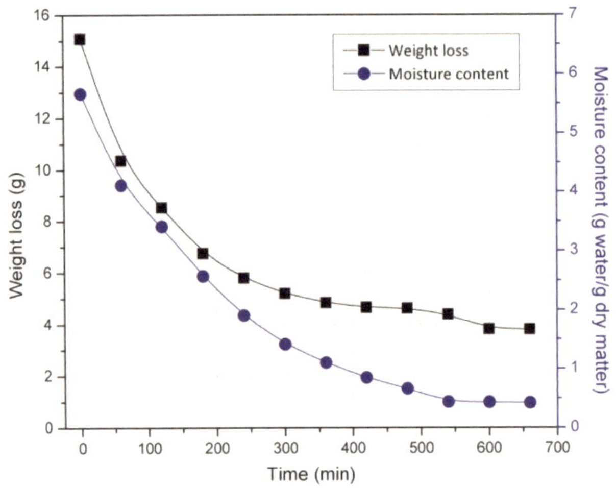

Figure 8.
Weight loss and moisture content depending on the drying time for Cymbopogon *leaves. Source: Own elaboration.*

final moisture content between 0.843 g water/g dry matter and 1.204 g water/g dry matter. Reaching a final drying rate between 0.643 g water/g matter dry minute and 1.024 g water/g matter dry minute, the final weight of the sample was 2.7 g (see **Figure** 7).

3.2.3 Cymbopogon

As can be seen in **Figure 8**, the drying kinetics of the *Cymbopogon* sample stabilized in 660 min. The initial moisture content was 5.66 g water/g dry matter. The final moisture content was between 0.421 and 0.435. **Figure 9** shows that the highest drying rate was 0.026 g water/g dry matter min.

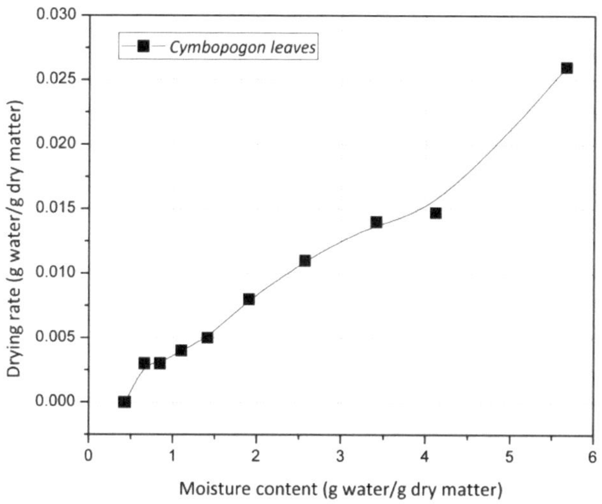

Figure 9.
Drying rate as function of moisture content for a Cymbopogon *leaves. Source: Own elaboration.*

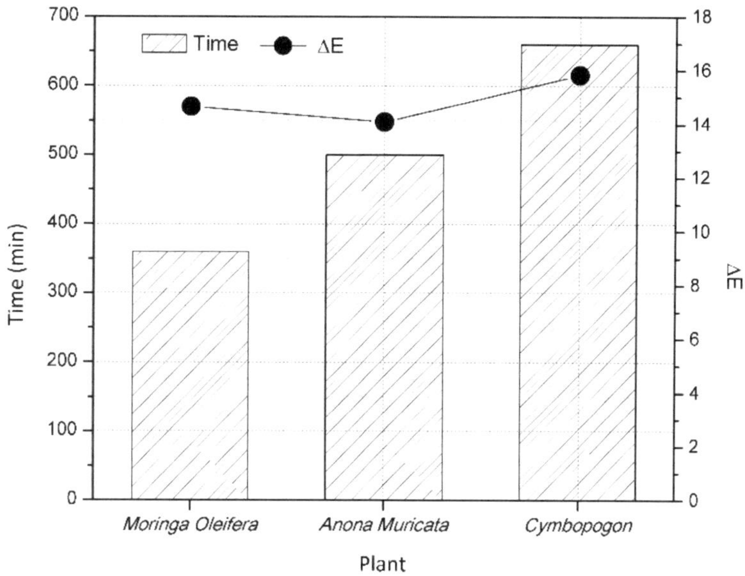

Figure 10.
Relationship between final drying time and ΔE in dehydrated leaves. Source: Own elaboration.

3.3 Colorimetric analysis

Color is a primary characteristic perceived by the consumer of a product and plays an important role in food. The color of food is often an indication of the nutrients it contains. In addition, the color of a processed product is expected to be as similar as possible to fresh ones [45].

Table 2 shows the values obtained from L*, a*, and b*, in the medicinal plants studied in fresh and dry.

An insignificant variation in the parameters L*, a*, and b* obtained in the plants studied before and after the drying process can be seen in **Table 2**. This can be corroborated in **Figure 10**, in which ΔE is analyzed. The main reason for this color preservation is the protection against solar irradiation of the solar tunnel dryer, so dehydrated plants are kept more similar to fresh ones. For this reason, it is preferable to dry in a closed and controlled environment, which is consistent with studies of Helvaci et al. [46].

It is important to note that in all cases the values of L*, a*, and b* decreased; this means that as the temperature increased (the three sections of the drying chamber reached up to 50.7°C). There was a tendency toward gray colors, a decrease in red colors (more brown), and an increase in yellow colors; these results agree with data reported by Bhardwaj et al. [47].

3.4 Computational analysis

Computational simulation is performed by using of a free software called Energy2D [48]. Simulation consists of temperature dependence upon time at dif-

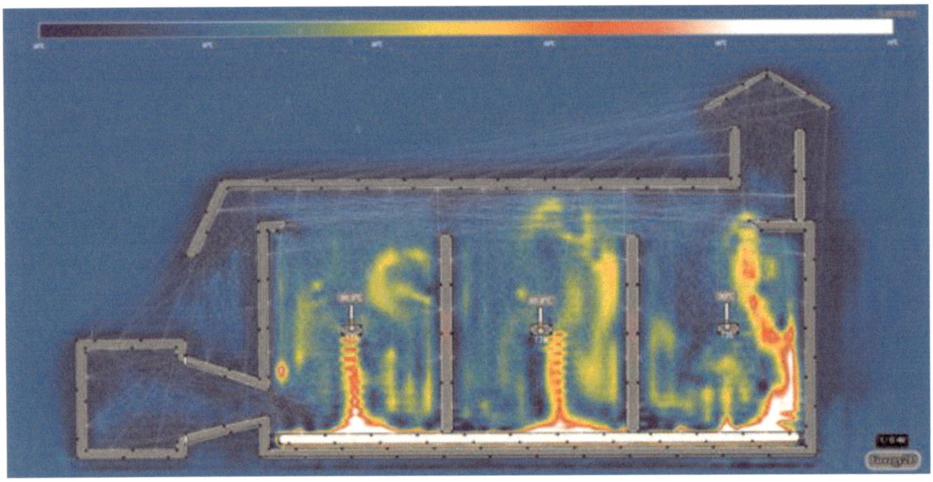

Figure 11.
Thermal behavior simulation of the dehydrator. Source: Own elaboration.

L*, a*, b* values and drying	**Medicinal plants**		
	Moringa oleifera	***Cymbopogon***	***Annona muricata***
Fresh leaves L*, a*, b* values	 L: 50.28 a: −6.76 b: 35.75	 L: 50.2 a: −7.87 b: 21.81	 L: 45.53 a: −8.41 b: 27.58
Dehydration with tunnel dryer L*, a*, b* values	 L: 40.4 a: −2.61 b: 25.75	 L: 41.27 a: 1.85 b: 13.14	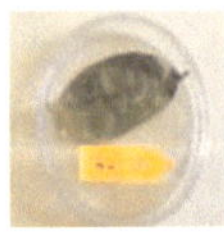 L: 37.02 a: −5.24 b: 16.81

Source: Own elaboration.

Table 2.
Values obtained from L, a*, b* (fresh and dried) in the medicinal plants studied.*

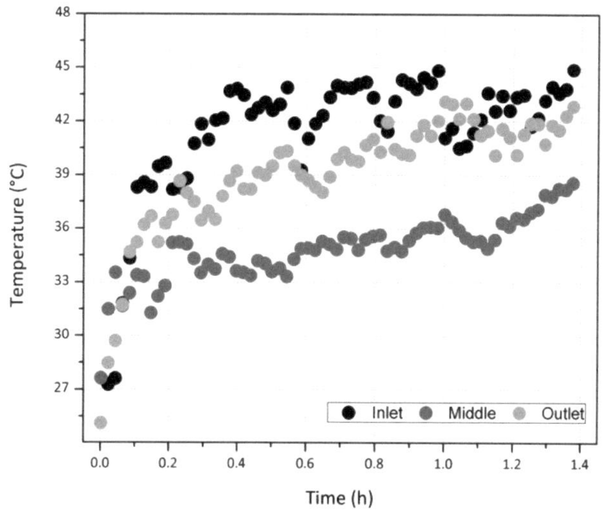

Figure 12.
Graph of thermal behavior simulation of the dehydrator. Source: Own elaboration.

Annona		
Model	**Coefficients and fit parameter**	**Value**
Modified page	k	0.0116
	n	0.5685
	R^2	0.9941
	RMSE	0.0199
	X^2	0.0005
Cymbopogon		
Model	Coefficients and fit parameter	Value
Modified page	k	0.0369
	n	0.7545
	R^2	0.9988
	RMSE	0.0092
	X^2	0.0001
Moringa		
Model	Coefficients and fit parameter	Value
Page	k	0.0067
	n	1.5616
	R^2	0.9987
	RMSE	0.0119
	X^2	0.0001
Source: Own elaboration.		

Table 3.
Results of regression analysis of the best adjusted models.

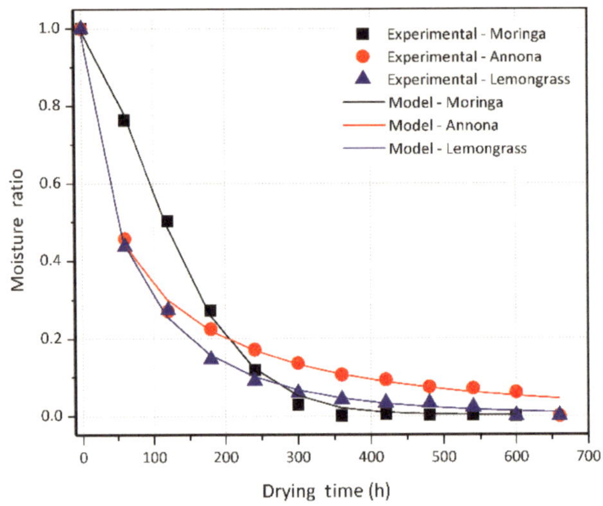

Figure 13.
Experimental versus mathematical model. Source: Own elaboration.

ferent zones. For the temperature evolution across time, thermal conduction and convection simulation mode was selected. For accurate results, also the thermal and optical properties of each material were chosen. The complete procedure of simulation can be found in Ref. [49]. In **Figure 11**, the thermal behavior of the dehydrator is shown and can be distinguished into three zones: left, middle, and right. Maximum reached values of temperature sensors, at the left, middle, and right zones were 45.2, 38.8, and 44.7°C, respectively, while its averages were about 41.4, 34.9, and 39.2°C, correspondingly. The temporal fluctuation of temperature is depicted in **Figure 12**.

The temperatures reached in the dryer with the hybridization of solar technologies remained stable during most of the day, with a variation between 45 and 50°C (**Figures 3** and **12**), which, as reported in the literature reviewed, are optimal to prevent the loss of important properties in vegetables and medicinal plants [46, 50].

3.5 Mathematical modeling

3.5.1 Adjusting experimental drying curves

The drying kinetics obtained experimentally in the medicinal plants were adjusted using the nine drying models mentioned in **Table 1**. **Table 3** presents the model that best fit for each drying condition. These models had the highest R^2 values as shown in **Table 2**.

Estimates of the statistical parameter for all experimental conditions showed that R^2 values ranged from 0.991 to 0.9988, indicating a good fit [34]. Therefore, these models adequately predict the selected plant drying kinetics. From the data, Modified Page model was the most suitable for describing drying processes of *Annona* and *Cymbopogon* drying conditions and Page for *Moringa* (see **Figure 13**).

4. Conclusion

Moringa oleifera, *Cymbopogon*, and *Annona muricata* have antioxidant, anticancer, anti-inflammatory, antidiabetic, antimicrobial, and many more positive effects in the human body. The drying kinetics of this plants were analyzed using a constructed indirect solar dryer type tunnel. It was possible to obtain optimum temperatures for drying (45–50°C), and drying times were significantly reduced by keeping these temperatures constant for a longer time throughout the day.

The quality of the plants at the end of drying was confirmed when obtaining average a_w values of 4.4. The average final moisture content of the three medicinal plants varied between 0.8 and 1.5 g water/g dry matter. The moisture ratio was fitted to nine thin-layer drying mathematical models to select a suitable drying curve. The Modified Page and Page models showed the best fit to the experimental results. The time required to reach the equilibrium moisture content in all experiments varied between 250 and 350 min. The superior efficiency of the dryer allows a reduction of the important drying costs. This technology would allow agricultural producers to reduce costs while contributing to the improvement of the environment.

Author details

Margarita Castillo Téllez[1]*, Beatriz Castillo Téllez[2], José Andrés Alanís Navarro[3], Juan Carlos Ovando Sierra[1] and Gerardo A. Mejia Pérez[2]

1 Facultad de Ingeniería, Universidad Autónoma de Campeche, Campus V, Unidad Habitacional Siglo XXIII por Avenida Ing. Humberto Lanz Cárdenas, Campeche, Mexico

2 Centro Universitario del Norte, Universidad de Guadalajara, Colotlán, Jalisco, México

3 Energy Department, Polythecnic University of Guerrero's State, Puente Campuzano, Guerrero, México

*Address all correspondence to: mcastill@uacam.mx

References

[1] Pérez Ortega G, González Trujano E. Plantas medicinales contra la ansiedad. Universidad del Zulia. 2015; **22**(3):68-75

[2] Sofowora A, Ogunbodede E, Onayade A. The role and place of medicinal plants in the strategies for disease prevention. African Journal of Traditional, Complementary, and Alternative Medicines. 2013;**10**(5): 210-229

[3] WHO Global Report on Traditional and Complementary Medicine 2019. Geneva: World Health Organization; 2019. Licence: CC BY-NC-SA 3.0 IGO

[4] Rastogi RP, Dhawan BN. Anticancer and antiviral activities in Indian medicinal plants: A review. Drug Development Research. 1990;**19**(1):1-12

[5] Surya S, Salam AD, Tomy DV, Carla B, Kumar RA, Sunil C. Diabetes mellitus and medicinal plants: A review. Asian Pacific Journal of Tropical Disease. 2014;**4**(5):337-347

[6] Abd Kadir SL, Yaakob H, Mohamed Zulkifli R. Potential anti-dengue medicinal plants: A review. Journal of Natural Medicines. 2013;**67**(4):677-689

[7] Nantia EA, Moundipa PF, Monsees TK, Carreau S. Les plantes médicinales dans le traitement de l'infertilité chez le mâle: Mise au point. Andrologie. 2009;**19**(3):148-158

[8] Marshall E. Health and wealth from medicinal aromatic plants. FAO; 2011. pp. 1-73

[9] Johnson RK et al. Dietary sugars intake and cardiovascular health: A scientific statement from the American Heart Association. Circulation. 2009; **120**(11):1011-1020

[10] Zhu DY, Bai DL, Tang XC. Recent studies on traditional Chinese medicinal plants. Drug Development Research. 1996;**39**(2):147-157

[11] Hasani-Ranjbar S, Jouyandeh Z, Abdollahi M. A systematic review of anti-obesity medicinal plants: An update (Provisional abstract)," Database of Abstracts of Reviews of Effects. 2013;**1**:28

[12] McCarty MF. ACE inhibition may decrease diabetes risk by boosting the impact of bradykinin on adipocytes. Medical Hypotheses. 2003;**60**(6): 779-783

[13] Gleye C, Laurens A, Hocquemiller R, Olivier L, Laurent S, Cavé A. Cohibins A and B, acetogenins from roots of *Annona Muricata*. Phytochemistry. 1997;**44**(8):1541-1545

[14] Rady I et al. Anticancer properties of Graviola (*Annona muricata*): A comprehensive mechanistic review. Oxidative Medicine and Cellular Longevity. Hindawi; 2018. pp. 39. Article ID: 1826170

[15] Adefegha SA, Oyeleye SI, Oboh G. Distribution of phenolic contents, antidiabetic potentials, antihypertensive properties, and antioxidative effects of Soursop (*Annona muricata* L.) fruit parts in vitro. Biochemistry Research International. 2015;**2015**

[16] Saroj K, Mukherjee PK, Pal M. Studies on some psychopharmacological actions of Moringa. 1996;**10**:402-405

[17] Perumal S, Klaus B. Antioxidant properties of various solvent extracts of total phenolic constituents from three different agroclimatic origins of drumstick tree (*Moringa oleifera* Lam.) Leaves. Journal of Agricultural and Food Chemistry. 2003;**51**(8):2144-2155

[18] Castillo TM, Castillo TB, Viviana MME, Carlos OSJ. "Technical and

experimental study of the solar dehydration of the moringa leaf and its potential integration to the sustainable agricultural industry," European Journal of Sustainable Development. 2018;7(3): 65-73

[19] Charles DJ. Antioxidant Properties of Spices, Herbs and Other Sources. Vol. 4. Springer; 2013. pp. 1-610

[20] qiang Zheng G, Kenney PM, Lam LKT. Potential anticarcinogenic natural products isolated from lemongrass oil and galanga root oil. Journal of Agricultural and Food Chemistry. 1993;**41**(2):153-156

[21] Aldawsari HM, Badr-Eldin SM, Labib GS, El-Kamel AH. Design and formulation of a topical hydrogel integrating lemongrass-loaded nanosponges with an enhanced antifungal effect: In vitro/in vivo evaluation. International Journal of Nanomedicine. 2015;**10**:893-902

[22] Bidinotto LT, Costa CARA, Salvadori DMF, Costa M, Rodrigues MAM, Barbisan LF. Protective effects of lemongrass (*Cymbopogon citratus* STAPF) essential oil on DNA damage and carcinogenesis in female Balb/C mice. Journal of Applied Toxicology. 2011;**31**(6):536-544

[23] Lee HJ, Jeong HS, Kim DJ, Noh YH, Yuk DY, Hong JT. Inhibitory effect of citral on NO production by suppression of iNOS expression and NF-κB activation in RAW264.7 cells. Archives of Pharmacal Research. 2008;**31**(3): 342-349

[24] Viuda-Martos M et al. Chemical composition and antioxidant and anti-listeria activities of essential oils obtained from some Egyptian plants. Journal of Agricultural and Food Chemistry. 2010;**58**(16):9063-9070

[25] Avila-Sosa R, Palou E, Jiménez Munguía MT, Nevárez-Moorillón GV, Navarro Cruz AR, López-Malo A. Antifungal activity by vapor contact of essential oils added to amaranth, chitosan, or starch edible films. International Journal of Food Microbiology. 2012;**153**(1–2):66-72

[26] Monge JN, Méndez-Estrada VH. Durango (México) y Costa Rica: dos maneras contrastantes de ver la educación a distancia (Durango (Mexico) and Costa Rica: Two contrasting views of distance). Rev de Educ Distancia. 2008;**28**(1):1-20

[27] Calixto JB. Efficacy, safety, quality control, marketing and regulatory guidelines for herbal medicines (phytotherapeutic agents). Brazilian Journal of Medical and Biological Research. 2000;**33**(2):179-189

[28] Cuellar-Nuñez ML, Luzardo-Ocampo I, Campos-Vega R, Gallegos-Corona MA, González de Mejía E, Loarca-Piña G. Physicochemical and nutraceutical properties of moringa (*Moringa oleifera*) leaves and their effects in an in vivo AOM/DSS-induced colorectal carcinogenesis model. Food Research International. 2018;**105**: 159-168

[29] Ani E, Amove J, Igbabul B. Physicochemical, microbiological , sensory properties and storage stability of plant-based yoghurt produced from bambaranut, soybean and *Moringa oleifera* seed milks. American Journal of Food and Nutrition. 2018;**6**(4): 115-125

[30] Luardini MA, Asi N, Garner M. Ecolinguistics of Ethno-Medicinal Plants of the Dayak Ngaju Community. Language Science; 2019

[31] T S, Al-Ismaili AM, Janitha Jeewantha LH, Al-Habsi NA. Effect of solar drying methods on color kinetics and texture of dates. Food and Bioproducts Processing. 2019;**116**: 227-239

[32] Guiné Raquel PF, Barroca Maria J. Effect of drying treatments on texture and color of vegetables (pumpkin and green pepper). 2013;**90**:58-63

[33] Shi Q, Zheng Y, Zhao Y. Mathematical modeling on thin-layer heat pump drying of yacon (*Smallanthus sonchifolius*) slices. Energy Conversion and Management. 2013;**71**:208-216

[34] Doymaz I, Smail O. Drying characteristics of sweet cherry. Food and Bioproducts Processing. 2011;**89**(1): 31-38

[35] Tunde-Akintunde TY. Mathematical modeling of sun and solar drying of chilli pepper. Renewable Energy. 2011; **36**(8):2139-2145

[36] Page GE. Factors influencing the maximum rates of air drying shelled corn in thin layes. Pardue University; 1949

[37] Diamante LM, Munro PA. Mathematical modelling of the thin layer solar drying of sweet potato slices. Solar Energy. 1993;**51**(4):271-276

[38] Pavis S, Henderson SM. Grain drying: Temperature effect on drying coefficients. Journal of Agricultural Engineering Research. 1961;**6**:169-174

[39] Togrul IT, Pehlivan D. Mathematical modelling of solar drying of apricots in thin layers. Journal of Food Engineering. 2002;**55**(3):209-216

[40] Koua KB, Fassinou WF, Gbaha P, Toure S. Mathematical modelling of the thin layer solar drying of banana, mango and cassava. Energy. 2009;**34**(10): 1594-1602

[41] Sharaf-Eldeen YI, Blaisdell JL, Hamdy MY. A model for ear corn drying. Transactions of ASAE. 1980; **23**(5):1261-1265

[42] Wang CY, Singh RP. A single layer drying equation for rough rice. American Society of Agricultural Engineers Paper no. 78–3001; 1978

[43] Midilli A, Kucuk H, Yapar Z. A new model for single-layer drying. Drying Technology. 2002;**20**(7):1503-1513

[44] Jin Y, Tang J, Sablani SS. Food component influence on water activity of low-moisture powders at elevated temperatures in connection with pathogen control. Lebensmittel-Wissenschaft+[ie und] Technologie. 2019;**112**:108257

[45] Gonçalves EM, Pinheiro J, Abreu M, Brandão TRS, Silva CLM. Modelling the kinetics of peroxidase inactivation, colour and texture changes of pumpkin (*Cucurbita maxima* L.) during blanching. Journal of Food Engineering. 2007

[46] Helvaci HU, Menon A, Aydemir LY, Korel F, Akkurt GG. Drying of olive leaves in a geothermal dryer and determination of quality parameters of dried product. Energy Procedia. 2019; **161**(2018):108-114

[47] Bhardwaj AK, Kumar R, Chauhan R. Experimental investigation of the performance of a novel solar dryer for drying medicinal plants in Western Himalayan region. Solar Energy;**177, 2019**(2018):395-407

[48] Xie C. Interactive heat transfer simulations for everyone. Physics Teacher. 2012;**50**(4):237-240

[49] Alanís Navarro JA, Castillo Téllez M, Martínez MAR, Silvar GP, Tejeda FCM. Computational thermal analysis of a double slope solar still using Energy2D. Desalination and Water Treatment. 2019;**151**:26-33

[50] Müller J, Heindl A. Drying of medicinal plants. Wild Relatives of Cultivated Plants in India; 2006. pp. 165-176

Chapter 2

Postharvest Treatment of Tropical Fruits Pineapple (*Ananas comosus*), Mamey (*Mammea americana*), and Banana (*Musa paradisiaca*) by Means of a Solar Dryer Designed

Italo Pedro Bello Moreira, Edgar Ruperto Macías Ganchozo, Xavier Enrique Anchundia Muentes, Celio Danilo Bravo Moreira, Manuel Eduardo Anchundia Muentes, Hebert Edison Vera Delgado and Carlos Eduardo Anchundia Betancourt

Abstract

The objective of this research was to know the useful life of dehydrated tropical fruits based on a solar dryer designed and developed under the conditions of Calceta, Bolívar Canton of the Province of Manabí, Ecuador. The physical and chemical characteristics exhibited during the radiation dehydration process were satisfactory, in fresh pineapple from 86.36% low humidity to 21.07%, from 0.67% protein to 2.45%, and from 2.05% fiber to 3.73%; in mamey from 79.30 to 21.07%, from 0.41 to 2.55%, and from 2.50 to 4.94%; and in bananas with from 80.22 to 10.35%, from 1.27 to 2.14%, and from 0.88 to 2.42. Microbiological analyses determined the life span of the products estimated at 174, 106, and 109 days, respectively, in pineapple, mamey, and banana. As for the attributes measured with the 1–5 scale of sensory evaluation, the mean treatments of their attributes such as color, sweetness, appearance, and taste were demonstrated where bananas present better color attributes with 4.38; 4.58, sweetness; 4.58, texture;4.68, appearance; and 4.75, flavor. Where significant diffraction can be determined relative to the calculated value $p > 0.05$ of <0.0001, the R^2 statistic in pineapple indicates 48.0814% variability in decreasing moisture pineapple (DMP), and its correlation coefficient is equal to 0.693408; the R^2 statistic in mamey indicates 55.6423% variability in decreasing moisture mamey (DMM), and its correlation coefficient is equal to 0.745938; and finally the R^2 statistic in banana indicates 56.339% variability in decreasing moisture banana (DMB), and its correlation coefficient is equal to 0.750593, indicating a moderately strong relationship between variables in all cases.

IntechOpen

Keywords: absorption, convective multiflash drying process (CMDF), fruit dehydration, fruit postharvest, solar energy, water activity

1. Introduction

In the Bolivar Canton of the city of Calceta of the Province of Manabí, Ecuador, the fruits pineapple (*Ananas comosus* (L.) Merr.), mamey (*Mammea americana* L.), and banana (*Musa paradisiaca* L.) are exuberant but due to high sugar and acid content have a limited shelf life and in many cases are wasted. Also, because these characteristics are easily adapted to conservation technologies, one of them is fundamentally dehydrated since they retain much of their taste, color, consistency, and appearance for long periods (see [1, 2]).

Sunlight from before and so far today serves to perform fruit drying as postharvest handling or conservation activity that, when purchased with other activities for the same purpose, such as chemical treatments, refrigeration, canned pre-eminence with operation, and collection even with solar drying, results in a decrease in energy expenditure spending and could therefore reach areas where other energy conveniences would not or even do not exist (see [3]).

In fruits the water present is 80.00% of its weight, for its early microbial decomposition is a definitive element. A conventional dehydrated fruit of its quality in terms of its organoleptic characteristics is lower than that of the fresh fruits from which it comes, affecting color, texture, and other peculiarities. With hot air, dehydrated fruits acquire levels of water activity (a_w) ranging from 0.6 to 0.8; these levels retain their sensory properties, showing good firmness to microbial attacks (see [4]).

Convective multiflash drying (CMFD) is a matter that is obtained after being used to obtain crispy fruits, and it is an option to the freeze-drying process (see [5]).

The industrial-scale study, a simple convective solar drying process of pineapples such as a round or circular economy tactic for countries in progress is specified, which is paramount to the tactic offered. It built a manual solar dryer that runs in indirect heat mode; it was also changed to improve its gain. The three main elements were raised that affect the convective drying process, the drying time (270 min, 480 min), the intensity of the solar radiation (650 W•m^{-21}, 100 W•m^{-2}), and the thickness of the cut (6–8 mm, 12–14 mm) (see [6]).

According to these instances, the aim of the research was to design and develop an empirical and practical distribution for solar dryer and drying tests to conserve the fruits and extend the shelf life, as well as organoleptic and nutritional particularities with samples of pineapple, mamey, and banana fruits in the city of Calceta, Bolívar Canton of the Province of Manabí, Ecuador.

2. Climate data collection

The research work was carried out from February to April 2013, in the school avenue of Calceta, Bolívar Canton, Province of Manabí, Ecuador, on the consecutive lines: South latitude 00′34.22′, west latitude 80 or 10′09.2′, average altitude of 22 masl, with a relative humidity of 90.00%, a temperature of 32.8°C, and a wind speed of 8 m.s^{-1}, until the end of November; the respective humidity is 86.7%, the temperature is 26°C, and the speed 1.8 m.s^{-1} (see [7]).

3. Design and construction of solar dryer

A dryer is used in heating the air by a 1 mm solar flat panel, insulated drying chamber equipped with stove where air is released for exhaust purposes. **Figure 1** describes the landscape of empirical or experimental distribution. The rustic surface of the solar collector is 0.5 × 0.5 × 1 m with a height of 0.7 m. The solar air heater is based on a folded suction plate with a dark bluish-look dyed uve (V) representation wood, as regards the exclusion of corresponding spaces on the sides as well as on the upper; the glass sheet was sealed with silicone-based material, diagonal to 15o, towards a dorsal part of the receiver (collector); a perforation was worked to carry out ventilation activities in the manner of air currents in other instances the side part, a lami was located a lami na (FV) photovoltaic to be able to understand the photoelectric effects that come from solar radiation, so with the receiver (collector) to dehydrate the fruits were stained bone white in all its accesses, covered with aluminum plate; the entrance had five ships distributed at a distance of 15 cm between each of them. The container was made of an aluminum-based wire mesh and glued to the frame inside the drying chamber. The collector outlet air enters the drying chamber at the bottom, immediately flowing into the upward orientation using the drying material. The camera was insulated from all sides except the top, the camera was tested with a fireplace for exhaust air, and the height of the fireplace was 0.25 m. These are the aspects that contemplate the construction of the solar dehydrator for the fruits (see [7]) (**Figures 2** and **3**).

3.1 System dimensioning

3.1.1 Collector area calculation

Busy strip (1.8 × 1.0) m. 1.80 m^2; Global radiation on average per day 1353 W•m^{-2}. Lots of temporary global units of solar radiation were calculated by manipulating a solar energy meter (Tenmar TM 207) with an accuracy of

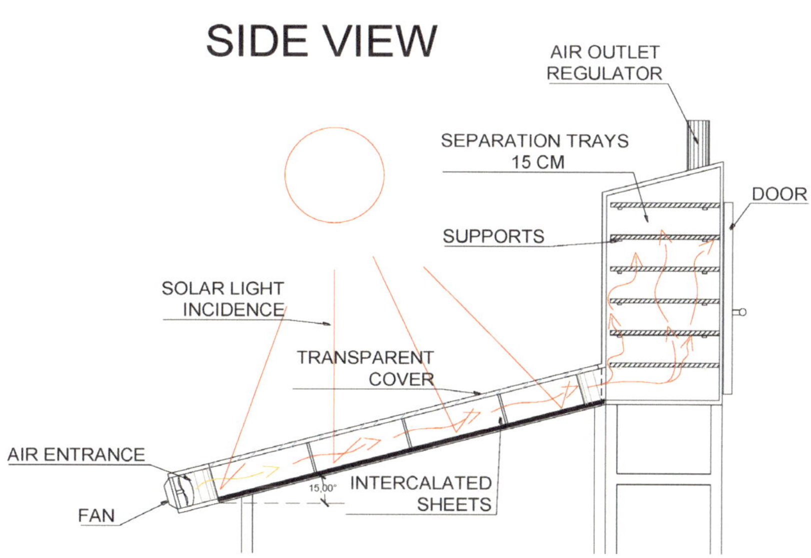

Figure 1.
Schematic view of experimental setup with solar flat plate collector.

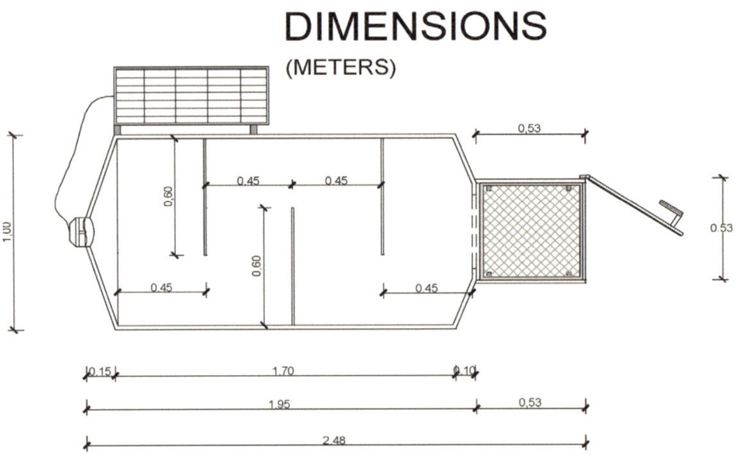

Figure 2.
End-floor view.

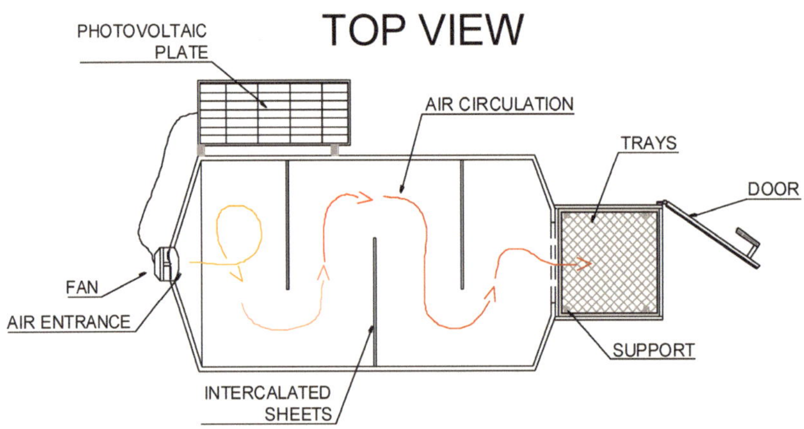

Figure 3.
Pair heating spray for process.

10 W•m^{-2}. Temperature readings were recognized every hour from 8:00 am to 6:00 pm. RTD were established at the inlet and outlet of the collector (Tin, Tout), outdoors to calculate the ambient temperature (Tamb) (see [8]).

3.1.2 Drying chamber

The volume of the drying chamber was established (oven example cabinet); consecutively, the average density of the fruits to be dried was 200 Kg.m^{-3}, with a mass of 4 kg; 4/200 to 0.02 m^3, being ten times more, a value close to 0 was obtained, 20 m^3.

3.1.3 Sterilization procedure

The matter is formed with preparations such as tools and components which I discuss the following:

The tools handled in the test consisted of sterilization at 180°C for 60 min according to the Medical Research Council; the chopping of the fruits was done in a sterilized part in advance rinsed with neutral soap; to quickly immerse the fruits in water with $C_6H_8O_6$ ascorbic acid to prevent oxidation and then fit into the dryer, start the test by placing the dehydrator; feel this part of the dryer, facing north, so that the collector takes the sun rays east to west. Three repetitions are executed for each fruit which are delayed from 3 to 5 days for each repetition. At the end of each day, the samples are wrapped in foil and sealed tightly and stored.

3.1.4 Statistical analysis of the values by treatments

Statistical analysis was contemplated with a complete or random design with therapizations in each of the treatments; the units are practices or experimental comprising of 4 kg of dried fruit. These results are tabulated making use of Statgraphics software and InfoStat™ 5.1TM in terms of linear regression and variance (ANOVA). To identify significant differences in treatments, as well as statistical significance for all comparisons, $p < 0.05$ was used. The Tukey multirange test was used to compare mean values of treatments.

4. Results

4.1 Bromatological analysis of fruits evaluated

At the bromatology laboratories of the Agricultural Polytechnic School in Manabí Manuel Félix López (ESPAM MFL) and CE.SE. C.CA Unibersidad Laica Eloy Alfaro de Manabí (ULEAM), Manta, Ecuador, analysis of moisture, ash, proteins and fiber is conducted, and 250 g was manipulated for each sample, published in **Table 1**.

The percentage of humidity in pineapple decreased from 86.36 to 21.14%, in mamey from 79.30 to 21.07%, and banana from 80.22 to 10.35%; ashes amount in pineapple from 0.44 to 1.09%; in mamey from 0.25 to 2.66%, and in bananas from 1.12 to 2.80% indicating that in the latter, it is higher than the previous fruits.

For pineapple protein amounts from 0.67 to 2.45%; for mamey from 0.41 to 2.55%, and for banana from 1.27 to 2.14%; in this case the three fruits increased their amounts, and finally the amount of fiber in pineapple is from 2.05 to 3.63%, in

Fruit state	Parameters	Method	Unity	Pineapple	Mamey	Banana
Fresh fruit	Moisture	INEN 864	%	86.36	79.30	80.22
	Ash	INEN 467	%	0.44	0.25	1.12
	protein	PEE/SECECCA/QC/15	%	0.67	0.41	1.27
	Fiber	PEE/SECECCA/QC/03	%	2.05	2.50	0.88
Dehydrated fruit	Moisture	INEN 864	%	21.14	21.07	10.35
	Ash	INEN 467	%	1.09	2.66	2.80
	protein	PEE/SECECCA/QC/15	%	2.45	2.55	2.14
	Fiber	PEE/SECECCA/QC/03	%	3.63	4.94	2.42

Table 1.
Bromatological analysis on fresh and dehydrated fruits.

mamey from 2.50 to 4.94%, and in banana from 0.88 to 2.42% of the same way augments, respectively.

4.2 System efficiency in each of the fruits

Figure 4 and **Table 2** show the moisture extracted against the emission of the radiation; according to this we manage to determine that the molecular structure of the banana facilitates the extraction more accelerated or rapidly with respect to moisture, thus following the pineapple and mamey in which it can be proven that the equipment is more efficient for banana fruit.

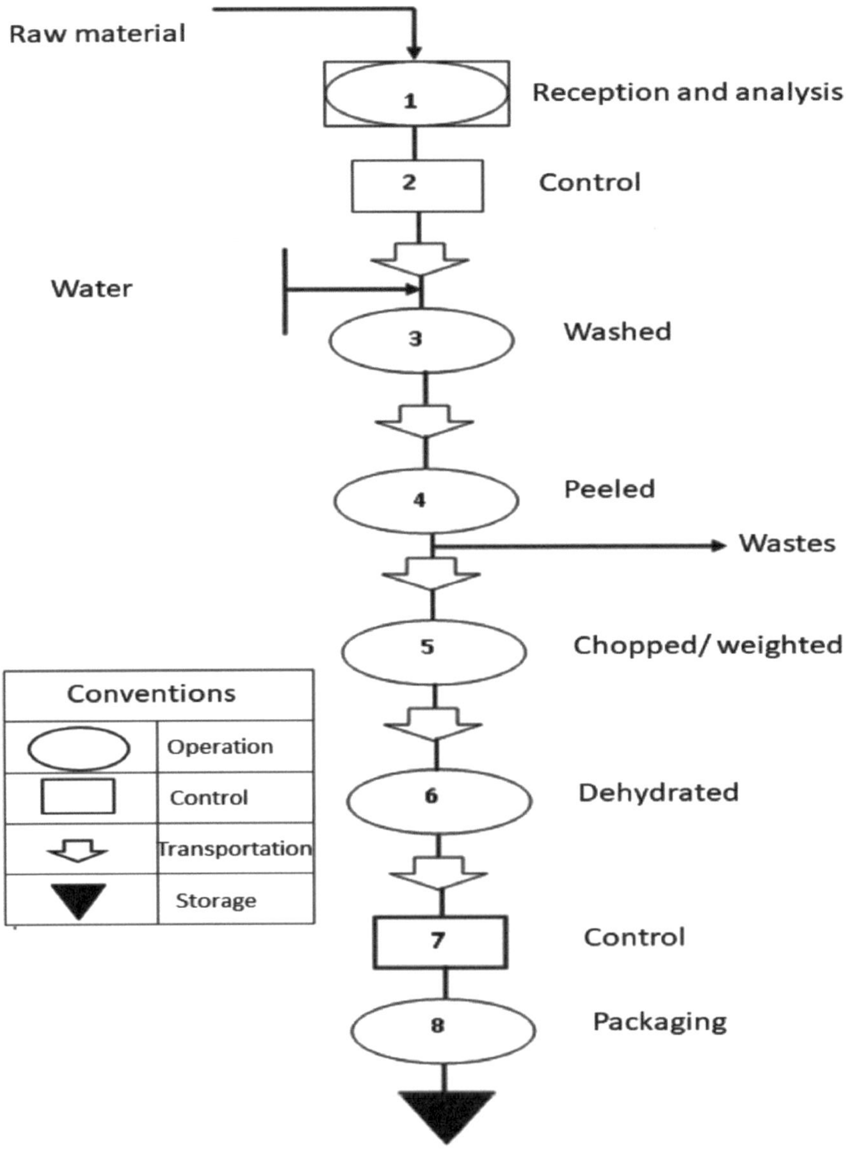

Figure 4.
Process flow chart of dried pineapple, mamey, and banana fruits.

Solar radiation ($W \bullet m^{-2}$)	Pineapple (% H)	Mamey (% H)	Banana (% H)
300	45.61883	46.54318	8.205
350	53.41883	50.89206	27.606
400	61.21883	55.24094	47.007
450	69.01883	59.58982	66.408
500	76.81883	63.9387	85.809
550	84.61883	68.28758	105.21
600	92.41883	72.63646	124.611
650	100.21883	76.98534	144.012
700	108.01883	81.33422	163.413
750	115.81883	85.6831	182.814
800	123.61883	90.03198	202.215
850	131.41883	94.38086	221.616
900	139.21883	98.72974	241.017

Table 2.
System efficiencies vs. dehydrated fruits.

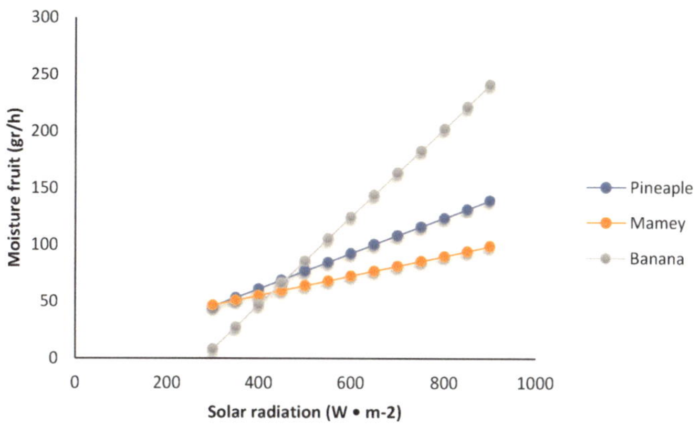

Figure 5.
Differentiation of water vapor contents as absolute humidity according to solar radiation.

Figure 5 shows the moisture variation as a function of solar radiation in the evaluated fruits, pineapple, mamey, and banana, respectively.

4.3 Microorganism tests to learn about fruit shelf life

In a given time of 15 and 30 days for the purpose of concerning the fruit drying activities assessed, each repetition is presented with their respective microbiological examinations with units of measurement in $CFU.g^{-1}$ which in turn were compared with Ecuadorian Standardization Service (INEN) standards with their details in maximum limits allowed; in other instances the microorganisms found in the samples requested from the microbiology laboratories of the Agricultural Polytechnic School of Manabí Manuel Félix López (ESPAM MFL) are multiplied exponentially, of mathematical type making use of exponents, i.e., the fickle x (unit of measure (d)) suppressed day presents the lifetime data for which the following are assessed: pineapple 174, mamey 106, and banana 109 (**Figure 6**).

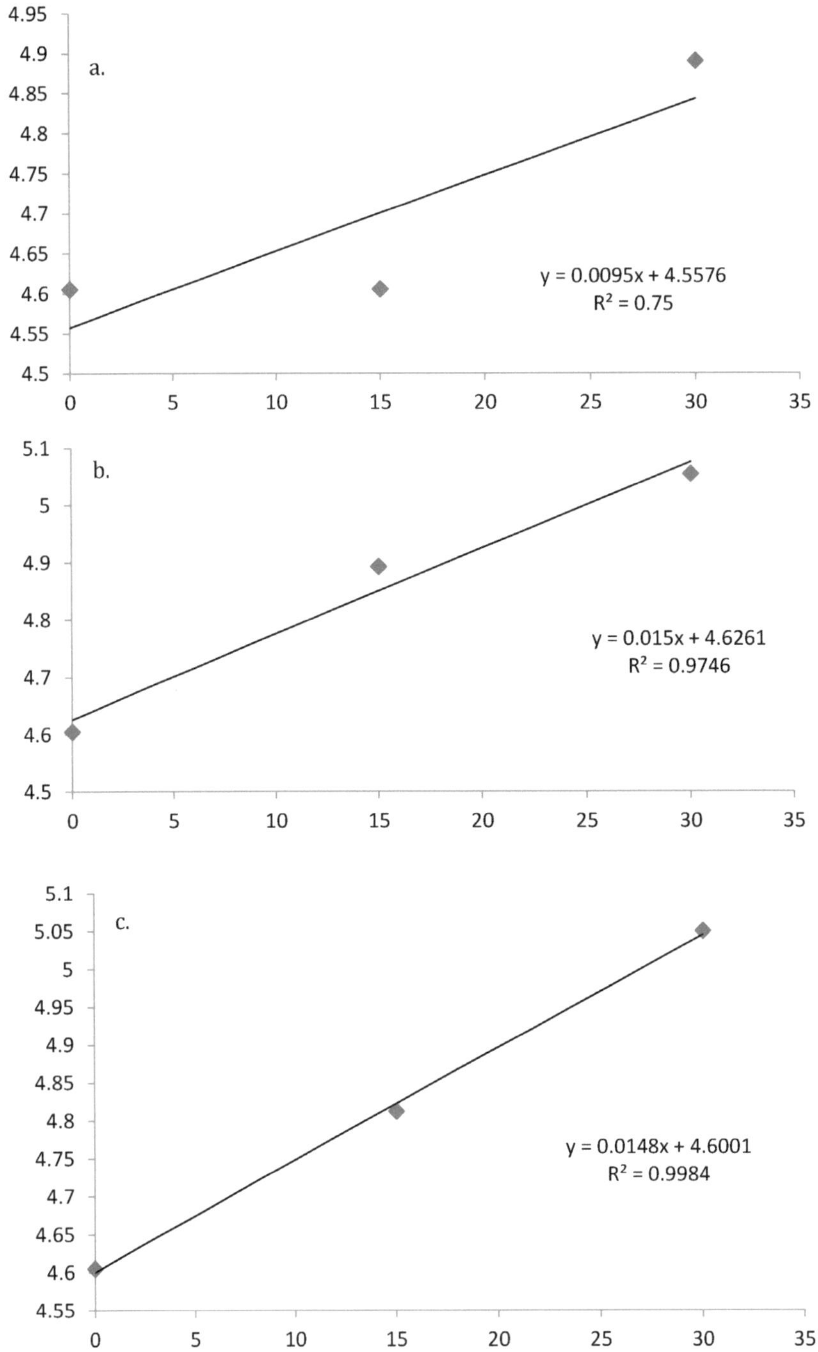

Figure 6.
Shelf life for dehydrated fruits. (a) Pineapple; (b) mamey; (c) banana.

4.4 Sensory evaluation with scale 1–5

The attributes evaluated were texture, sweetness, aspect, color, and flavor; these were calculated on the basis of the InfoStat software, allowing the most relevant characteristics of the attributes to be in their respective order for dehydrated fruits of pineapple, mamey, and banana, respectively (**Figures** 7 and 8, **Tables** 3–5).

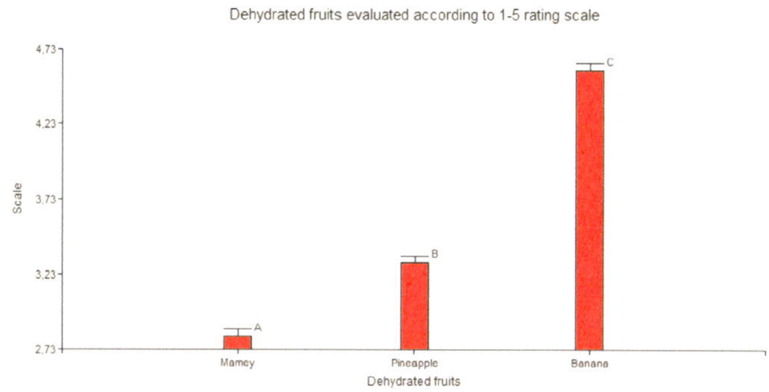

Figure 7.
The valuation scale in dehydrated fruits.

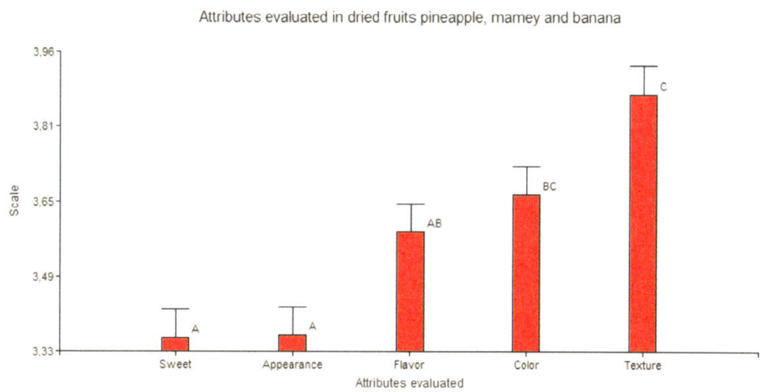

Figure 8.
Evaluation of dehydrated fruit acrations depending on the scale score.

Variable	N	R^2	R^2Aj	CV
Scale	300	0.76	0.75	12.87

Table 3.
Analysis of variance.

F.V.	SC	gl	CM	F	p-value
Model	193.09	14	13.79	65.32	<0.0001
Factor A	168.56	2	84.28	399.17	<0.0001
Factor B	11.12	4	2.78	13.16	<0.0001
Factor A*Factor B	13.42	8	1.68	7.94	<0.0001
Error	60.17	285	0.21		
Total	253.27	299			

Table 4.
Variance analysis (SC type I).

error: 0.2111 gl: 285

Factor A	**Factor B**	**Half**	**n**	**E.E.**					
Mamey	Sweet	2.48	20	0.10	A				
Mamey	Appearance	2.51	20	0.10	A				
Mamey	Flavor	2.53	20	0.10	A				
Pineapple	Appearance	2.91	20	0.10	A	B			
Pineapple	Sweet	3.03	20	0.10		B	C		
Mamey	Color	3.21	20	0.10		B	C	D	
Mamey	Texture	3.35	20	0.10		B	C	D	
Pineapple	Color	3.40	20	0.10		B	C	D	
Pineapple	Flavor	3.48	20	0.10			C	D	
Pineapple	Texture	3.70	20	0.10				D	
Banana	Color	4.38	20	0.10					E
Banana	Sweet	4.58	20	0.10					E
Banana	Texture	4.58	20	0.10					E
Banana	Appearance	4.68	20	0.10					E
Banana	Flavor	4.75	20	0.10					E

Means with a common letter are not significantly different ($p > 0.05$).

Table 5.
Test: Tukey Alpha = 0.05 DMS = 0.49402.

4.5 Relationship and recoil tests concerning the access temperature in dryer with the receiver access sheet (collector), as well as the radiation with the low amounts of water as absolute humidity in the fruits of pineapple, mamey, and banana

Based on p-value in ANOVA calculated less than 0.01, therefore, there is a significant relationship from the statistical level, between the drying T°C and the plate T°C with 99% confidence level.

R^2 indicates the percentage (%) variation of the response variable that explains its relationship to one or more predictor variables; it can be said that the higher the R^2, the better the arrangement of the model to the obtained data.

So according to this premise, the model exposes 61.3763% variability in drying T°C, and the correlation coefficient is equal to 0.78343; therefore, it indicates a moderately strong relationship between the variables, and the standard error evaluation shows in the calculation that the standard deviation of the residuals is 2.51359.

Regression exam/linear pattern $Y = a + b * x$.

Dependent variable (VD): drying T°C

Independent variable (VI): plate T°C

The results expose the output, and it conforms to the linear model to refer to the relationship between the drying T°C with the plate T°C.

The equation of the adjusted model below is

$$Drying\ T°C = 20.0773 + 0.275953 * plate\ T°C \tag{1}$$

4.5.1 Regression examination between drying temperature and receiver temperature (collector)

Depending on p-value in ANOVA less than 0.01, therefore, there is a significant statistical relationship between the drying T°C and the T°C of the collector with a level of 99% confidence.

R^2 indicates that the model exposes 59.672% variability in drying T°C, and the correlation coefficient is equal to 0.772477, therefore indicating a moderately strong relationship between the variables (**Figure 9**).

Regression exam/linear pattern $Y = a + b * x$

Dependent variable (VD): drying T°C

Independent variable (VI): collector T°C

The results expose the output, and it conforms to the linear model to refer to the relationship between the drying TOC and the collector TOC.

The equation of the adjusted model below is

$$Drying\ T°C = 23.2088 + 0.269216 * Collector\ T°C \tag{2}$$

4.5.2 Pineapple fruit and your regression exam

Depending on p-value in the ANOVA less than 0.01, there is a significant statistical relationship between DMP and SRP with a 99% level of trust.

R^2 indicates that the model exposes 48.0814% variability in DMP. The correlation coefficient is equal to 0.693408; therefore, it indicates a moderately strong relationship between the variables, and the standard error evaluation of the sample in the calculation shows that the standard deviation of the residuals is 23.4888 (**Figure 9a**).

Linear pattern $Y = a + b * x$

Dependent variable (VD): decreasing moisture pineapple (DMP)

Independent variable (VI): solar radiation pineapple (SRP)

The effects of linear model adjustment refer to the relationship between DMP and SRP.

The equation of the adjusted model below is

$$DMP = -1.18117 + 0.156 * SRP \tag{3}$$

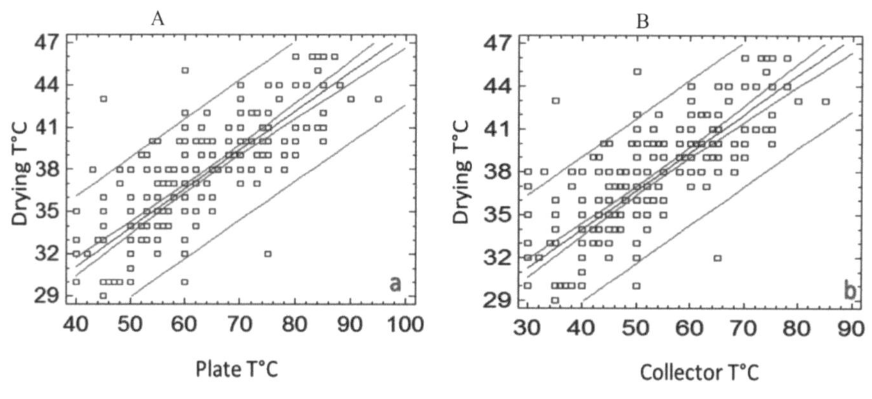

Figure 9.
Drying inlet with plate temperature (a) and receiver temperature (collector) (b), where the adjusted model is displayed.

4.5.3 Mamey fruit and your regression analysis exam

Depending on the p-value in ANOVA calculation which is less than 0.01, there is a significant statistical relationship between DMM and SRM with a confidence level of 99%. The R^2 states according to model that there is 55.6423% variability in DMM, the correlation coefficient is equal to 0.745938; therefore, it shows a moderately dynamic relationship between the variables, and the standard error evaluation shows that the standard deviation of the residuals is 12.3989 (**Figure 9b**), as well as the average absolute error (AAE) of 10.7287.

Linear pattern $Y = a + b * x$

Dependent variable (VD): decreasing moisture mamey (DMM)

Independent variable (VI): solar radiation mamey (SRM)

The effects of linear pattern adjustment to refer to the dependency between DMM and SRM.

The equation of the adjusted model below is

$$DMM = 20.4499 + 0.0869776 * SRM \tag{4}$$

4.5.4 Banana fruit and your regression analysis exam

Based on ANOVA's p-value in calculation less than 0.01, there is a significant statistical ratio of DMB to SRB with a confidence level of 99%, while R^2 exposes according to model that there is 56.339% variability in DMB; in other instances the correlation coefficient the evaluation is equal to 0.750593; therefore, it teaches the coordinately dynamic relationship between the variables and finally the standard evaluation error presenting a value of 35.5063 (**Figure 10c**). The mean absolute error (AAE) is 24.6024.

Linear pattern $Y = a + b * x$

Dependent variable (VD): decreasing moisture banana (DMB)

Independent variable (VI): solar radiation banana (SRB)

The effects of linear pattern adjustment refer to the dependency between DMB and SRB.

The equation of the adjusted model below is

$$DMB = -108.201 + 0.38802 * SRB \tag{5}$$

4.6 Examination of the process by psychometric spread

The leaf concerning psychometric dehydration consents to be able to observe two autonomous methodologies, i.e., at one point, there is a sensitive air heated in the receiver or collector, in which air entering the apparatus is heated at the cost of emissions of solar radiation sponsoring moisture content firmly, and consecutively, in other instances the air develops moisture by vaporizing water in the fruits of pineapple, mamey, and banana as a result of cooling. As an example, in the first process of dehydration of banana fruit, we assume subsequent identifications with their means: air temperature, 26°C; first relative humidity, 72%; air temperature at the dryer inlet as well as the leak in the receiver (collector), 49°C; air temperature in dryer escape, 35.7°C; air channel diameter, 0.08 m; speed of air at the entrance, 16.6 $m.s^{-1}$; and separate humidity of water, 172.43 g.

4.7 Through these ante-laid characterizations, the movement of the mass fluids was established using the following equations

$$W = \frac{\pi * d^2}{4} * V * p \tag{6}$$

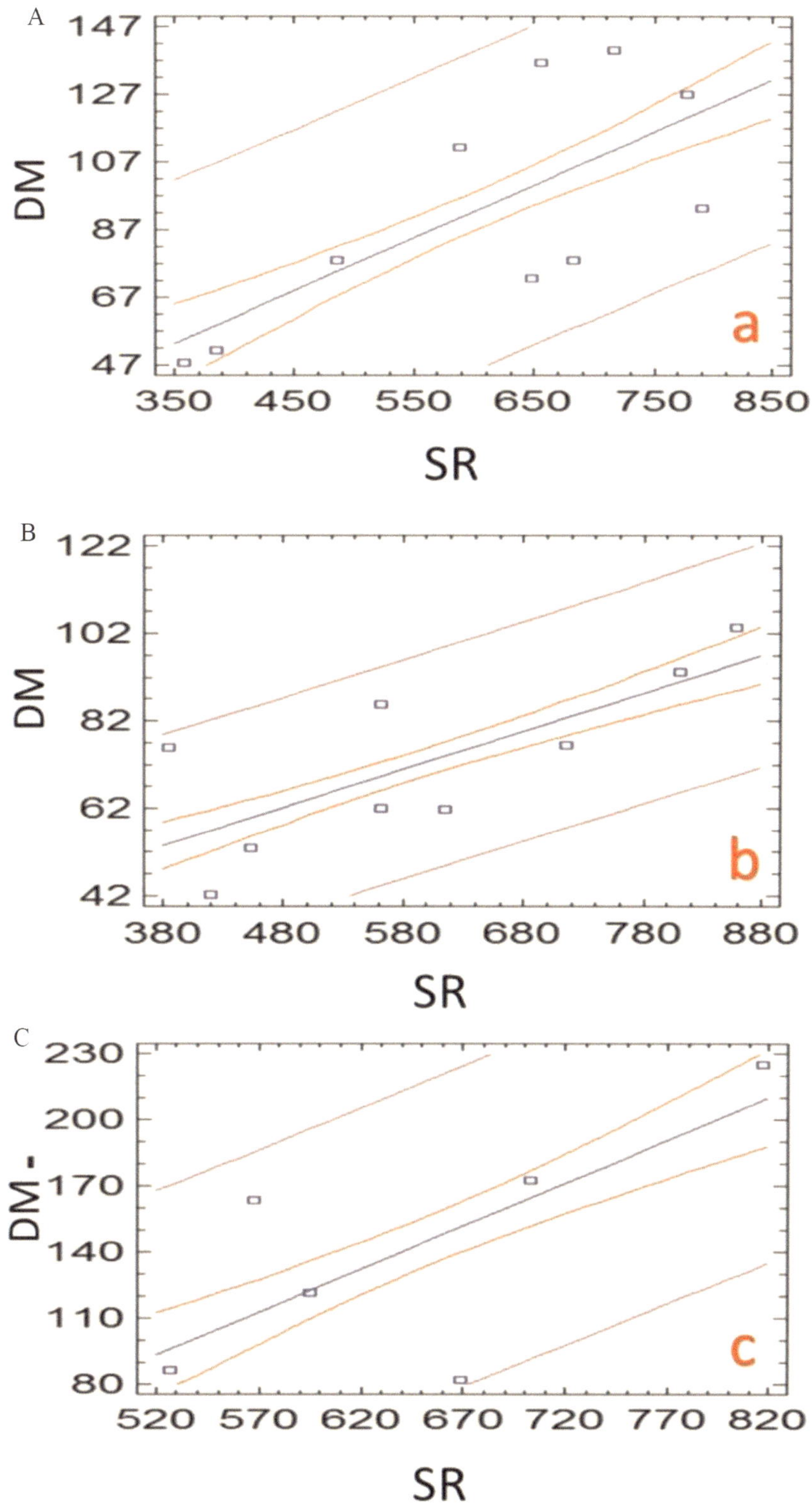

Figure 10.
Pattern provided towards low humidity in fruits dehydrated, (a) pineapple, (b) mamey, and (c) banana, with respect to solar radiation.

where d is the diameter of the air channel and V is the speed of air as well as the air density, resulting in the value of 0.083 $kg.s^{-1}$. Consecutively after these values and identifying the suppressed humidity of the fruits, the portion of water absorbed in kg of air entering the dryer in the staged phase is established by using Eq. (7):

$$d = \frac{172.43\, g\ water}{1801\, k\ dried\ air} = 0.095\frac{g}{k} \tag{7}$$

In **Figure 11**, you can observe the thermodynamic moments of sites 1, 2, and 3 provided to the air and habitat, in the receiver leak (collector) as well as in the dryer leak; therefore, these identifications, as data taken from psychrometric chart software, manifest the temperature of the dry lamp(DB); relative humidity (RH); absolute humidity (AH); specific volume (Vol); enthalpy(Ent); steam pressure (VP); dew point(DP); and wet bulb temperature (WBT).

The case in process 1–2 was framed concerning reflective heating in the solar receiver (collector), while in other instances in the case corresponding to process

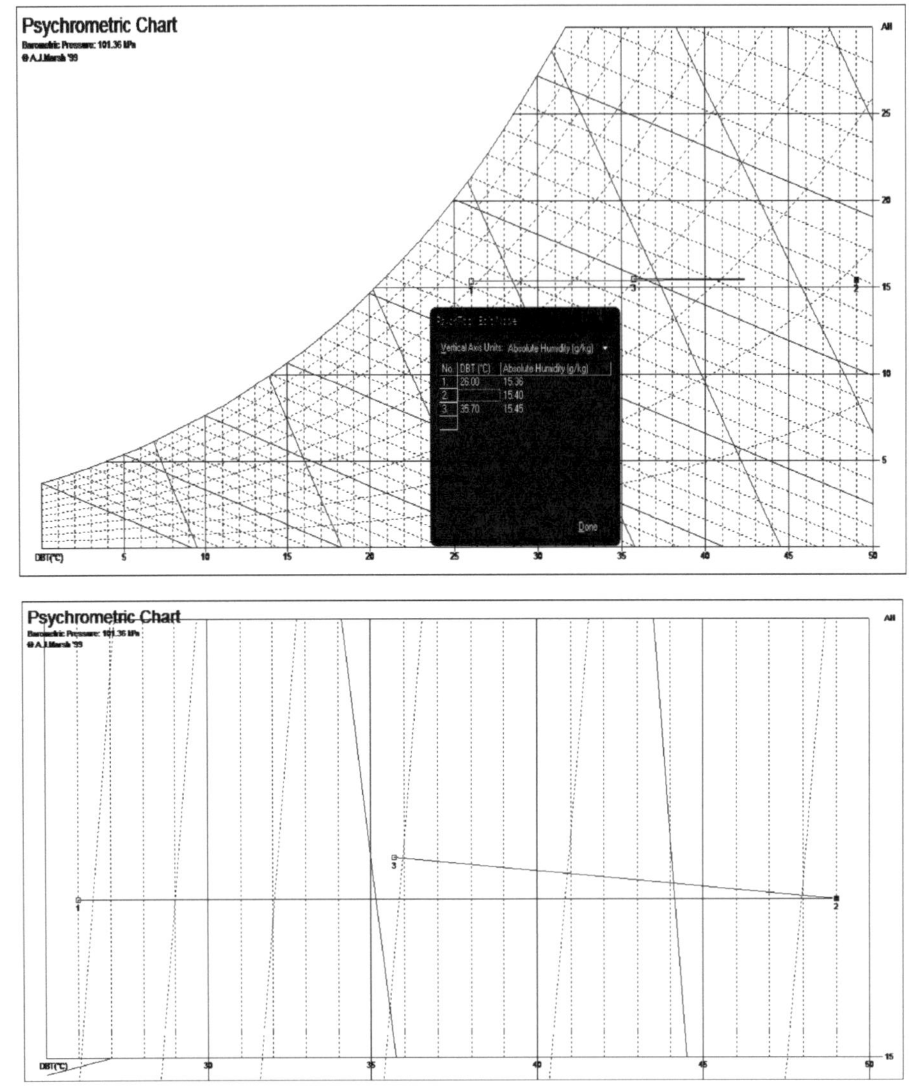

Figure 11.
Examination of the table in collector and solar dryer in the form of a psychometric chart.

2–3 in the inlet dryer, the humidification of the air and its convenient cooling were carried out through the activities of evaporation of water included in the fruits of pineapple, mamey, and banana.

5. Debate

In **Table 1**, we can disclose the percentage (%) of humidity above what was known after the dehydration activities: for pineapple with 86.36%, the moisture content is reduced by an average of X 2.14%, in mamey with 79.30%, by an average of X 21.07%, and for banana with a total of 80.22%, by an average of X 10.35%. The data are consistent with ante positioned knowledge executed by Almada (see [9]). There are also other intellectual citations covered in the mastery of the convective multiflash desiccation process (CMFD) (see [7–11]) are correlated in dried fruits, the inquiry as a study subject, for banana fruit were heated at 60°C by hot air as well as a vacuum pulsation was used, therefore consecutively in the drying of the fruit by medium of the convective drying and dizzying banana vaporization mixture, by CMFD, showed a moisture amount at 0.293 $g.g^{-1}$ (dry base) then in 3 h of process, the pattern of heat diagnosed in banana samples existed through the direction (conduction) and solar luminescence, so they were heated by approximately 60°C, large increases in independent or free water (banana moisture content at 76.00%), coinciding with the illustrations executed (see [10, 11]).

Subsequently in the convective multiflash drying process (CMFD) (about 135 min of process), the banana achieved a moisture amount in 0.29–0.01 $g.g^{-1}$; finally in the processing, the quantities were 0.276–0.015. The fruits with average amounts of moisture showed a water action ranging from 0.65, 0.85 to 0.90, while the mass of the fruits, due to dehydration, was noted with reduced amount of water and dominoes up to 60.21 g of solid in each repetition of 500 g of dough, mainly because of the water and native mechanisms are transferred to the osmotic procedure from extracellular areas, producing an ascent in the aroma of the fruit as well as well as the texture (see [11, 12]).

This work resolved to disinfect at 180°C for 60 min according to internationally determined rules such as the Medical Research Council; fruits such as banana amounts in 80.22% moisture in fresh state. Together, the validity of the dehydrator was select, assuming the similarity of the convective multiflash drying process (CMFD) method (see [4]) and the solar dryer. In both cases containing the use of solar radiation emission, these being at 300–900 $W{\bullet}m^{-2}$ (see [13]) the proportion of solar radiation previously and subsequently in the matter of dehydration within the process activities, the banana fruit had a reduction of 80.22–10.35%, exposing that subsequently it is being subjected to 180°C for 60 min a day with a total of 15–30 days. Its index was 0.12% concerning 10.35%, pointing to 0.4 times minimum amount in parallel in banana fruit.

The convective multiflash drying process (CMFD) is a method of efficient dehydration due to two important reasons. Firstly because during flash evaporation, it starts from moisture that encloses internally, so it is pull towards the surface of the fruits, making optimal the convective drying during the movement for heating purposes. Secondly we have to subsequently evaporate the flash, and the fruit is stothered to the surface area of it, making optimal the convective drying during the heating movement as well as later in the evaporation of the flash. The temperature of the fruit drops by 15–20°C, so it transports a relevant difference in the temperature between the hot air and the cooled fruit. We could therefore say that because of this, the fruit receives a better transfer of heat, see [7], a method similar to the procedure or technique assessed in this investigation as the object of study.

Having had on balance, all this provides more accurate and complete assessments of the duration as useful life of dehydrated fruits, in pineapple, mamey and banana individually, of 15–30 days of dehydration process in repetitions, the amount of the shelf life of banana fruit, by sample, see **Figure 5**, as linear regressions with adjusted model of the low humidity of banana fruit in this regard with radiation showed a duration of 109 days, the moisture content of pineapple was reduced from the amount or average riginaria by 86.36% to a final amount of 21.07%; mamey fluctuated between 79.3 and 21.13% and bananas fluctuated between 80.22 and 10.35%, proportionally. As well as, properties such as ash in pineapple at 0.44% had an increase with an average of 1.09%; 0.25% had an increase of 2.66%, and in banana fruit it fluctuated between 1.12 and 2.80%, exposing, in banana farmed fruits, data with more expensive values (data not exposed).

The protein content, as well as fiber, was known a characteristic increase significantly in the previous and subsequent periods in dehydration, is understood as a derivation of fresh pineapple fruit in 0.67% protein and fibers at 2.45% in its fresh phase, together, in dehydrated period, had an increase of 2.45 and 3.63%. The mamey fruit with 0.41% protein and fiber sin at 2.5% in its fresh period, in its dehydrated period, had an increase of 2.55 and 4.94%, proportionally, finally, in banana fruit with amounts of protein type by 1.27 and 0.88% in its fresh, dehydrated time, had an increase of 2.18 and 2.41%, correspondingly.

6. Termination

This work as an object of study in the research assessed the incidences of a solar dryer, which in turn was developed in the city of Calceta, in Bolívar Canton of the Province of Manabí, Ecuador, with the approach of drying to such a point of being able to talk about the phenomena concerning the dehydration of fruits such as pineapple, mamey, and banana. Therefore, trials were carried out as experiments finding, for example, that the organization of the molecular structure of the banana provides the facility for moisture content by means of solar radiation activity.

Then, followed by the pineapple fruit and mamey, as a result, this loss is compressing the moisture content of the valued fruits. The temperature of the drying air is the most important and effective component of several elements of the system during drying; air humidity, as well as air speed, is also a significant factor in optimizing the drying rate. Therefore, in these concerns, it is understood that the microbiological examination helps to establish the shelf life of dehydrated fruits, and it is more important to mention that from 15 to 30 days, repetitions in measured quantities CFU.g^{-1} have given legitimate maximum values, which is understood as a result of the proliferation of microorganisms in the fruits evaluated.

Acknowledgements

We thank the Unibersidad Laica Eloy Alfaro de Manabi (ULEAM), Ecuador, for being part of the university community as teachers and researchers, Secretariat of Higher Education, Science, Technology and Innovation (Senescyt), and Faculty of Agricultural Sciences of the Agroindustrial Engineering Career in turn that in some way or another allows us to universalize knowledge and transmit it to students.

Dedication

This chapter of the book is dedicated to God, who gives me day-to-day forality to make the daily activities, to my wife Emily Julissa Mendoza Cedeño and daughter

Hadassah Julissa Bello Mendoza for their spiritual support and words of encouragement to make diary, and to my esteemed fraternal friend Edgar Ruperto Macías Ganchozo whose idea and professional work could have made this chapter a reality for his knowledge.

Appendices and nomenclature

Roman symbols

E	solar radiation intensity ($W \bullet m^{-2}$)
A	surface area (m^2)
n	number of experiments
CFU	colony-forming unit
t	temperature
h	specific enthalpy (kJ/kg)
RH	relative humidity
AH	absolute humidity
Vol	specific volume
VP	vapor pressure
DP	dew point
WBT	temperature of wet bulb
R^2	percentage of variation of the variable
Eq.	equation
AAE	average absolute error

Greek symbols

τ	time, min

Subscripts

amb	ambient (at 25°C, 30 wt % relative humidity, 1.012 bar) (kJ/kg)
DM	dry matter
CMFD	convective multiflash drying process
W	water
D	the drying chamber of the drying process
DMP	decreasing moisture pineapple
SRP	solar radiation pineapple
DMB	decreasing moisture banana
SRB	solar radiation banana
DMM	decreasing moisture mamey
SRM	solar radiation mamey

Author details

Italo Pedro Bello Moreira*, Edgar Ruperto Macías Ganchozo,
Xavier Enrique Anchundia Muentes, Celio Danilo Bravo Moreira,
Manuel Eduardo Anchundia Muentes, Hebert Edison Vera Delgado
and Carlos Eduardo Anchundia Betancourt
Faculty of Agricultural Sciences, Career of Agroindustrial, Agricultural
and Environmental Engineering of the Laica University "Eloy Alfaro" of Manabi,
Manta, Ecuador

*Address all correspondence to: italop.bello@uleam.edu.ec

References

[1] Hernández RJ, Flores MF, Acosta OR, Barbosa PG. Secado solar de productos agrícolas. Caos Conciencia. 2014;**8**(1): 25-34. Available from: http://dci.uqroo.mx/RevistaCaos/2014_vol1/Secado_Solar.pdf

[2] Hernández RJ, Martínez VO, Quinto DP, Cuevas DJ, Acosta OR, Aguilar JO. Secado de chile habanero con energía solar. Revista Iberoamericana de Tecnología Postcosecha. 2010;**10**(2):120-127. Available from: http://www.redalyc.org/articulo.oa?id=81315091008

[3] Reinoso B, Diseño ES. Construcción de un secador experimental de hierbas aromáticas con el empleo de energía solar, Capacidad de 5 Kg [Thesis]. Sangolquí/ESPE, Ecuador: Universidad de Fuerzas Armadas; 2006

[4] Zotarelli M, Almeida PB, Borges LJ. A convective multi-flash drying process for producing dehydrated crispy fruits. Journal of Food Engineering. 2011; **108**(4):523-531. DOI: 10.1016/j.jfoodeng.2011.09.014

[5] García AI, Justinovich S, Angel L, Heredia T. Secadero solar forzado para productos agrícolas. Averma. 2015; **19**(2):21-28. Available from: http://www.asades.org.ar/Averma/Secadero%20solar%20forzado%20para%20productos%20agricolas

[6] Ochoa MCI. Red neuronal artificial en respuesta a predicciones de parámetros de transferencia de masa (pérdida de humedad y ganancia de sólidos) durante la deshidratación osmótica de frutas. Acta Agronomica. 2016;**65**(4):318-325. DOI: 10.15446/acag.v65n4.50382

[7] Azouma YO, Drigalski L, Jegla ZK, Reppich M, Turek V, Weiß M. Indirect convective solar drying process of pineapples as part of circular economy strategy. Energies. 2019;**12**(15):2841. DOI: 10.3390/en12152841

[8] INAMHI-Instituto Nacional de Meteorología e Hidrología de Ecuador. Geoinformación Hidrometeorológica de Manabí y Esmeraldas, Ecuador. Red de Estaciones Meteorológicas Convencionales. Geoinformación Hidrometeorológica de Manabí y Esmeraldas, [Internet]. 2016. Available from: http://www.serviciometeorologico.gob.ec [Accessed: 11 September 2019]

[9] UNESCO. Organización de la Naciones Unidas para la Educación, la Ciencias y la Cultura Guia de Uso, de Secadores Solares para frutas, Legumbres, Hortalizas, Plantas Medicinales y Carnes, Agencia Suiza para el Desarrollo y la Cooperación (Ministerio Suizo de Asuntos Exteriores) [Internet]; Paraguay, Asuncion; 2005. pp. 25-28. Available from: https://unesdoc.unesco.org/ark:/48223/pf0000156206

[10] Sivipaucar C, Curo H, Huancahuari E, Llantoy V, Valderrama A. Cálculo y Construcción de un Secador Solar por Convección Natural para el Secado de Plantas Medicinales no Tradicionales. Huarochirí, Peru: Cedit-Centro de Desarrollo e Investigación en Termofluidos; 2008. pp. 18-30. Available from: http://sisbib.unmsm.edu.pe/BibVirtual/Publicaciones/rev_cedit/2008_V03/pdf/ a03v3.pdf

[11] Gamboa D, Ibáñez D, Meléndez M, Paredes E, Siche R. Secado de lúcuma (*Pouteria obovata*) empleando la técnica de Ventana Refractante™. Scientia Agropecuaria. 2014;**5**(2):103-108. Available from: http://revistas.unitru.edu.pe/index.php/scientiaagrop/article/view/580/542

[12] Ortiz GS, Sánchez LL, Valdés RMP, Baena GD, Vallejo CFA. Efecto de la osmodeshidratación y secado en la retención de carotenos en fruto de zapallo. Acta Agronomica. 2008;**57**(4): 269-274. Available from: https://revistas.unal.edu.co/index.php/acta_agronomica/article/view/9265/9923

[13] CONELEC-Consejo Nacional de Electricidad. In: Tacuri FI, CIE Corporación para la Investigación Energética, editors. Atlas Solar del Ecuador con fines de Generación Eléctrica [Internet]. Quito, Ecuador; 2008. Available from: http://biblioteca.olade.org/opac-tmpl/Documentos/cg00041.pdf

Section 2

Convective Drying

Chapter 3

Convective Drying in the Multistage Shelf Dryers: Theoretical Bases and Practical Implementation

Artem Artyukhov, Nadiia Artyukhova, Ruslan Ostroha, Mykola Yukhymenko, Jozef Bocko and Jan Krmela

Abstract

The main advantages regarding the convective drying of the granular materials in the multistage dryers with sloping perforated shelves were represented. Peculiarities of the shelf dryers' hydrodynamics were shown in the research. Various hydrodynamic weighing modes were experimentally justified, and the relevant criteria equations were obtained. The results of investigations regarding the interphase heat and mass transfer were given; criteria dependencies, which predict heat and mass transfer coefficients in the shelf dryers, were proposed. A method to assess the efficiency of the dehydration process at the separate stages of the device and in the dryer, in general, was proposed. The algorithm to define the residence time of the granular material on the perforated shelf with a description of the author's software product for optimization calculation was shown. The shelf dryers' engineering calculation method was presented in this work. The original constructions of devices with various ways to control the residence time of the granular material that stays in their workspace were described. The testing results of the shelf dryer to dry granular materials, such as coarse- and fine-crystalline potassium chloride, sodium pyrosulfate, and iron and nickel powders, were demonstrated.

Keywords: convective drying, multistage shelf dryer, hydrodynamics, interphase heat and mass transfer, efficiency of dehydration, engineering calculation, industrial implementation

1. Introduction

Convective drying is one of the most effective methods for disperse material dehydration in the chemical, pharmaceutical, mining, and other industrial branches [1–6]. The direct contact of the drying agent with high-temperature potential and dried material enables intensively to remove the surface and adsorption-bound moisture [3]. That is why drying is the only method for industries to achieve the required quality of the product.

Although during drying the energy consumption is the lowest, the convective dryers are more often used to dry disperse materials thanks to other numerous advantages [3, 4]. Provision of the active hydrodynamic regime in such dryers helps intensify the process without reduction of the economic efficiency of their operation and has the following advantages [7–9]:

- Hydrodynamic stability of the process
- Increase of the relative motion velocity of the interacting phases
- The developed surface of the contacting phases
- Approximation of the hydrodynamic model of flows in the device to the ideal displacement model
- Reduction of the energy intensity of the process and metal intensity of devices

One should distinguish devices with various configurations of the weighted layer (fluidized bed, spouting, gravitational falling, vortex, etc.,) from the variety of the convective dryer constructions (described, e.g., in [7, 10]). Thanks to the developed surface of the phase contact, devices with hydrodynamic system are characterized with high intensity of the heat and mass transfer processes, lower non-energy costs, and have high specific productivity [7].

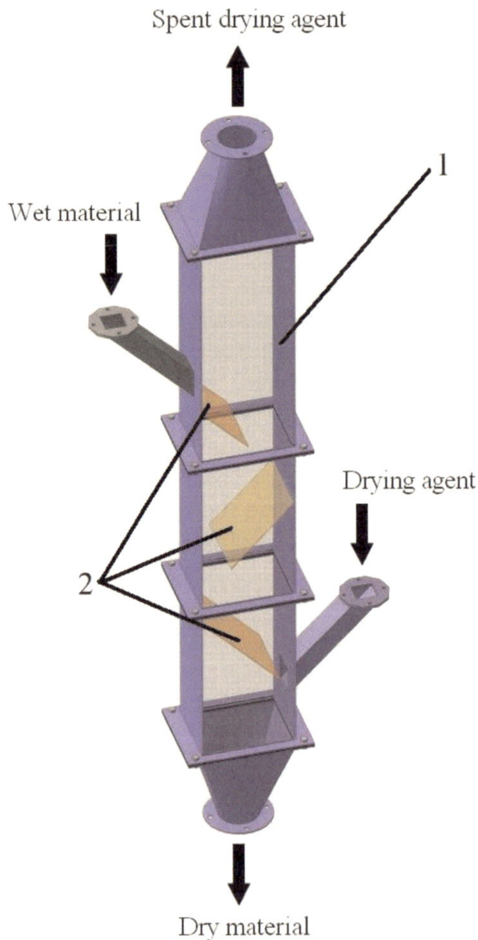

Figure 1.
Multistage shelf dryer: (1) case; (2) shelf.

Besides, under conditions of the cost increase to prepare and to transport the drying agent, the possibility of its repeated use in the drying process is fundamentally important. Therefore it is necessary maximum to use the thermal potential of the drying agent during every contact with disperse material. It can be achieved while using the multistage drying devices with vertical sectioning of the workspace by the perforated shelf elements [7, 8, 11].

A solution to the permanent residence of the dispersed material in the "active" zone can be found through implementation of the multistage shelf dryers with vertical sectioning of the workspace (**Figure 1**).

In such devices, the conditions for the differentiated heat treatment of materials can be created owing to the drying agent potential and peculiarities of each stage (shelf) construction. So, moving the perforated shelf to the wall of the device, we approach to the device with fluidized bed, and moving it away from the wall and freeing its workspace, we approach to the device with the free intersection, such as pneumatic transportation dryers. The shelves increase the residence time of the dried material particles, either poured downward of the device or carried out upward of it. The shelves increase the velocity and turbulence of the gas flow, create a vortex motion in their location places, and increase the contact intensity between phases. The free space between the ends of the shelves and the walls of the device does not require special flows from the upper shelf to the lower one. Changing the free area of the shelve perforation, their tilt angles, the distance from the end of the shelves to the walls of the device, the number of shelves, and the distance between them vertically, it is possible to influence the intensity of the contact phases and to create different hydrodynamic regimes to weigh particles of the material both on individual shelves and heightwise the device. Therefore, it is possible to carry out the drying process of wet materials and its pneumatic classification in one device in order to remove small dusty fractions from the mixture [7].

2. Theoretical basics

The necessity to determine these features is due to the fact that before constructing an industrial sample of the gravitational shelf dryer it is necessary to determine its optimal design. In this case, the optimization criterion is to ensure the minimum required residence time of the dispersed phase in the working space of the dryer, which will complete the drying process to a predetermined humidity indicator. It is important to observe the condition, under which the "hydrodynamic" residence time of the dispersed phase in the workspace of the device should be no less than the "thermodynamic" time (this parameter is determined by the kinetics of the moisture removal process from the dispersed phase). Therefore, in order to keep the integrity of the dispersed particles, the "hydrodynamic" time should not exceed the "thermodynamic" time by more than 5–10%. By adjusting the hydrodynamic properties of the flow, an optimal construction of the gravitational shelf dryer is achieved, which meets the requirements of the optimization criterion.

Thus, the optimization calculation of the dryer consists of three blocks: hydrodynamic calculation (calculation of the residence time of a particle on a stage), kinetic calculation (kinetic parameter of the moisture removal), and calculation of drying efficiency.

Initial data (**Figure 2**):

- Rate of gas flow, Q (m^3/s)
- Length of device, L (m)

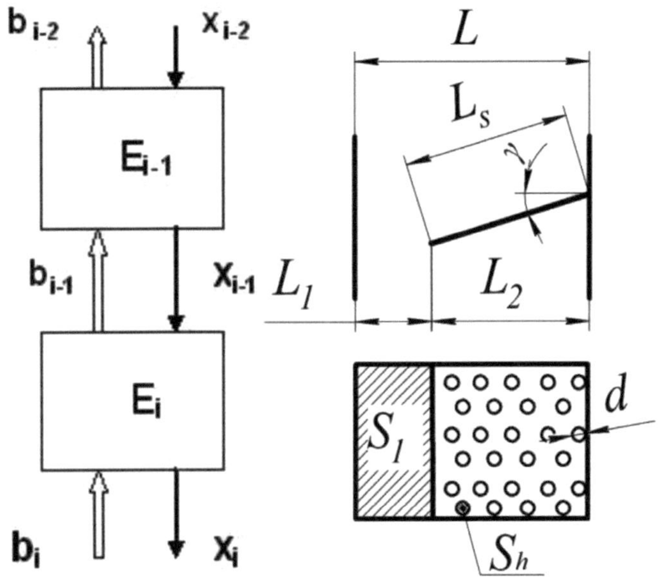

Figure 2.
A fragment of the calculation scheme for the multistage drying: left figure—change of flows' moisture: x, moisture of the disperse material; b, moisture of the drying agent, and right figure—construction of dryer's workspace.

- Overall width of device, h (m)
- Length of shelf, L_s (m)
- Degree of perforation (free area), δ
- Perforation hole diameter, d (m)
- The tilt angle of shelf, γ (°)
- The radius of the granule, r_{gr} (m)
- Granule density, ρ_{gr} (kg/m^3)
- Gas density, ρ_g (kg/m^3)
- Acceleration of gravity, g (m/s^2)
- Resistance coefficient, ξ
- Volumetric content of a dispersed phase in a two-phase flow, ψ
- The coefficient that takes into account the tightness of the flow, m
- Number of stages in dryer, i
- Moisture of the material i-stage of the dryer x (kg of water/kg of material)
- Moisture of the drying agent in i-stage of the dryer b (kg of water/kg of material)

2.1 Hydrodynamic calculation

Hole area on the shelf (horizontal position) (m^2)

$$S_h = \frac{\pi d^2}{4}. \quad (1)$$

The perforated area on the shelf (horizontal position of the shelf) (m^2)

$$\sum S_h = L_s \cdot h \cdot \delta. \quad (2)$$

Number of holes on the shelf

$$n = \frac{\sum S_h}{S_h}. \quad (3)$$

Area of outloading clearance (m^2)

$$S_1 = (L - L_s \cos\gamma) \cdot h. \quad (4)$$

Area of the gas passage holes in the shelf (inclined position of the shelf) (m^2)

$$S_2 = \frac{\pi d^2}{4} \cdot n \cdot \cos\gamma. \quad (5)$$

The relative area of outloading clearance

$$S_1^r = \frac{S_1}{S_1 + S_2}. \quad (6)$$

The relative area of the gas passage holes in the shelf

$$S_2^r = \frac{S_2}{S_1 + S_2}. \quad (7)$$

Rate of the gas flow in outloading clearance (m^3/s)

$$Q_1 = Q \cdot S_1^r. \quad (8)$$

Rate of the gas flow in holes in the shelf (m^3/s)

$$Q_2 = Q \cdot S_2^r. \quad (9)$$

Gas velocity in holes in the shelf (m/s)

$$V_{work} = {}^{Q_2}\!/_{S_2}. \quad (10)$$

Second critical velocity (m/s)

$$V_{cr} = 1.63 \cdot \sqrt{\frac{\rho_{dr} \cdot g \cdot r_{gr}}{\xi \cdot \rho_g}}. \quad (11)$$

Velocity difference (m/s)

$$\Delta V = V_{cr} - V_{work}. \quad (12)$$

Time of material residence on the shelf (free movement) (s)

$$\tau_f = \frac{L_s}{\Delta V \sin\gamma}. \tag{13}$$

The empirical function of the effect of compression on the residence time of the particle in the working space of the device

$$f_{e\tau}(\psi) = (1-\psi)^{-m}. \tag{14}$$

Time of material residence on the shelf (straitened movement) (s)

$$\tau_s = \frac{L_s \cdot f_{e\tau}(\psi)}{\Delta V \sin\gamma}. \tag{15}$$

The program Multistage Fluidizer® [12] used Hypertext Markup Language HTML, Cascading Style Sheet (CSS), and programing language JavaScript (including the library jQuery). HTML is presented as a tagging of web-based app, CSS page formatting. JavaScript is used to calculate and transfer data and to create animation and data validation effect. In the validation block of JavaScript, data accuracy is checked. In the block input info, the basic data field indices are accepted, and they are written to the object of input_information. In the block, calculation computations are carried out by Eqs. (1)–(15).

Index.html (**Figure 3**) is the main page of the web-based app. It is responsible for reflection of the main menu, for the main calculation of gas flow, and for jumping the other pages (an example of such page is presented in **Figure 4**), where the main dependencies between key features to calculate gas flow and resistance time of the material on the shelf are calculated and dependencies diagrams are formed.

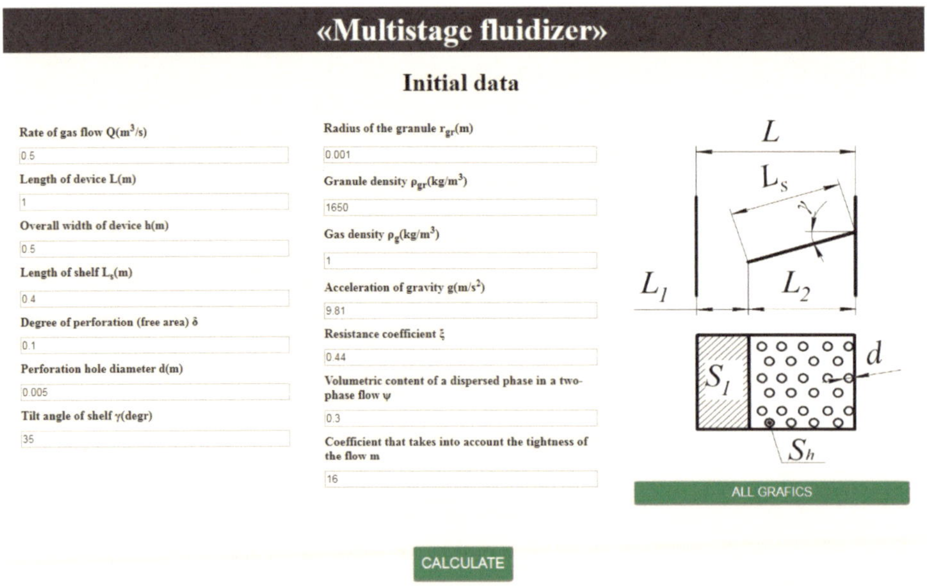

Figure 3.
The main page of the Multistage Fluidizer® software.

Influence of radius of the granule on the residence time of a particle

Rate of gas flow $Q(m^3/s)$	1
Length of device $L(m)$	0.8
Length of shelf $L_s(m)$	0.7
Overall width of device $h(m)$	1
Minimum radius of the granule $r_{grMin}(m)$	0.005
Maximum radius of the granule $r_{grMax}(m)$	0.05
Step of radius of the granule $\Delta r_{gr}(m)$	0.005
Degree of perforation (free area) δ	0.1
Perforation hole diameter $d(m)$	0.007
Tilt angle of shelf γ(degr)	15
Granule density $\rho_{gr}(kg/m^3)$	1650
Gas density $\rho_g(kg/m^3)$	1
Acceleration of gravity $g(m/s^2)$	9.81
Resistance coefficient ξ	0.44
Volumetric content of a dispersed phase in a two-phase flow ψ	0.3
Coefficient that takes into account the tightness of the flow m	16

CALCULATE

Figure 4.
Calculation page of various parameter impacts on the particle resistance time in the device.

Having inserted data, data validity is tested, that is, if all data is correct, after keystroke CALCULATE data is processed given the above formulas, and we receive the result in a form of a computation table (**Figure 5**).

After changes of indices l_f and l_s, it is possible to see how the animation appears after the pressing the "Show calculation" button (the example of distance length calculation, which particle takes on the shelf during the specified period of time, is shown in **Figure 6**). In order to create animation, functions clicker_lf and clicker_ls are implemented, respectively, for τ_f and τ_s animation.

In JavaScript, one uses libraries jQuery and table2excel, objects for data recording, methods .val() .append() to read and to insert indices to fields, methods

Calculation of the residence time of a particle on a stage

Hole area on the shelf (horizontal position), S_h (m^2)	0.00001963
Perforated area on the shelf (horizontal position of shelf), $\sum S_h$ (m^2)	0.02000
Number of holes on the shelf n	1019
Area of outloading clearance, S_1 (m^2)	0.3362
Area of the gas passage holes in the shelf (inclined position of shelf), S_2 (m^2)	0.01638
Relative area of outloading clearance S_1^r	0.9535
Relative area of the gas passage holes in the shelf S_2^r	0.04647
Rate of gas flow in outloading clearance, Q_1 (m^3/s)	0.4768
Rate of gas flow in holes in the shelf, Q_2^r (m^3/s)	0.02323
Gas velocity in holes in the shelf, V_{work} (m/s)	1.418
Second critical velocity, V_{cr} (m/s)	9.886
Velocity difference, ΔV (m/s)	8.468
Time of material residence on the shelf (free movement), τ_f (s)	0.08235
Empirical function of the effect of compression on the residence time of the particle in the working space of the device $f_{er}(\psi)$	300.9
Time of material residence on the shelf (straitened movement), τ_s (s)	24.78

Figure 5.
Results of calculation.

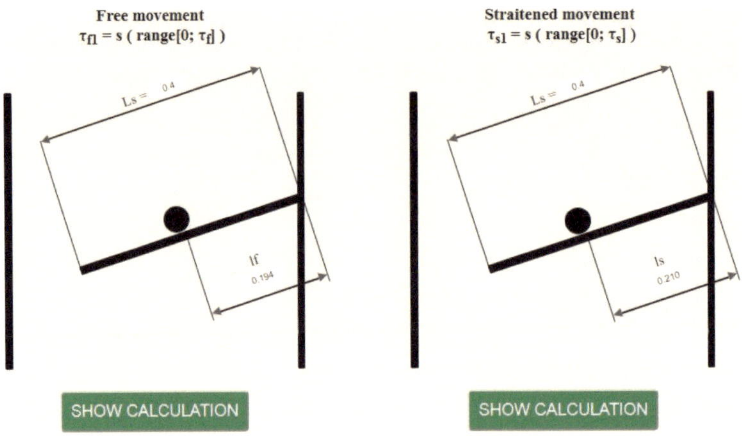

Figure 6.
Calculation of way length, which particle undergoes on the shelf during the specified period of time.

.removeClass() and .addClass() to delete and to add classes, and method .animate() for work with animation to create an animation effect for any digital CSS feature of the element.

2.2 Calculation of heat-mass transfer

In order to calculate the kinetic parameter of the moisture removal (the moisture-yielding capacity coefficient), let us use the following algorithm.

It is proposed [8] to use the following equation for calculation of β:

$$\frac{\Delta U_m}{\tau} = \beta \cdot F \cdot \left(b_{fin} - \frac{b_{fin} + b_{in}}{2} \right) \cdot \rho_g, \tag{16}$$

where b_{in}, b_{fin}, and ΔU_m are the initial and final humidity of the drying agent and the amount of the removed moisture from the material; F is the surface of the mass transfer, which depends on the dryer's effective operation due to the disperse material and the residence time of the material in the dryer.

In general, the criteria equation of the drying process can be written as follows:

$$Sh = A_1 \cdot Sc^n \cdot \mathrm{Re}^m, \tag{17}$$

where A_1 is the equation coefficient; $Sh = \frac{\beta \cdot d_{gre}}{D}$ is the Sherwood criterion; d_{gre} is the equivalent diameter of the particle (granule), m; $Sc = \frac{\nu}{D}$ is the Schmidt criterion; $\mathrm{Re} = \frac{V_{work} \cdot d_{gre}}{\nu}$ is the Reynolds criterion; D is the diffusion coefficient of the gas flow, m^2/s; ν is the kinematic viscosity coefficient of the gas flow, m^2/s; and m and n are the indicators of the equation stages, which are evaluated through the graphical dependency $Sh/Sc^{0.33} = f(\, Re \,)$, obtained from the experimental data.

2.3 Calculation of the kinetics and drying efficiency

The drying process effectiveness on the i-stage of the dryer is presented by the ratio of differences between the moisture contents of the disperse material before and after the drying $x_{i\text{-}1} - x_i$ to the maximum possible (theoretical) difference between the moisture contents on the stage $x_{i-1} - b_i$ and also in the form of the function of the kinetic parameter of the moisture transfer B_i, the residence time of the material on the stage τ_i, and the consumption ratio of the dispersed phase to the drying agent G_i^{-1} [8]:

$$E_i = \frac{\Delta x}{\Delta x_{\max}} = \frac{x_{i-1} - x_i}{x_{i-1} - b_i} = \frac{1 - \exp\left[-\beta_i \tau_i \left(1 + G_i^{-1}\right)\right]}{1 + G_i^{-1}}. \tag{18}$$

3. Visualization of results and discussion

3.1 Hydrodynamics

Some graphic dependencies are shown in **Figure** 7. The program receives two- and three-dimensional dependency graphs.

In general dependency diagrams, features for free and constraint motion of particles have one functional dependence. The particle resistance time has enough narrow diapason at every stage (shelf) in free motion regime and is calculated by second units. In the constraint motion regime of particles, the residence time is greatly increased at every stage. The abundant ratio of particles in the two-phase system has a definite impact on this index. That is why, while defining the optimum performance of the device, it is necessary to define the workspace size of the granulating or drying device to high accuracy.

The impact made by some constructive features of the shelf dryer during the residence time of the dispersed material (**Figure** 7) is shown below.

The change of the shelf tilt angle to the horizon affects the redistribution of the gravity components: enlargement of it leads to an increase of the gravity rolling component and vice versa. It should be mentioned that the tilt angle of the shelf may have a minimum value that complies with the natural slope angle of the material. As the tilt angle of the shelf decreases, the residence time of the dispersed material gradually increases. It leads to longer contact with the drying agent's flow.

Changing the gap between the edge of the shelf and the dryer's wall significantly influences the change of the residence time of the dispersed material on the shelf. If the gap increases, the contact time of the dispersed material with the drying agent will be reduced due to the decrease in the length of the material movement distance on the shelf. In this case, the operation of the rolling component of the dispersed material velocity lasts for a shorter period and at the end of the shelf is replaced by the full gravity. Thus, the material moves down, and only the ascending gas flow force resists its fall.

The analysis of the calculations regarding the effect, made by the free cross-sectional area of the shelf on the drying process efficiency, showed the following. Reducing the free cross-sectional area of the shelf leads to an increase of the drying agent's ascending motion velocity in the holes. In this case, the action of the drying agent's ascending flow slows down the progressive motion of the dispersed material on the shelf, compensating for the rolling component of its gravity. The pulse component of the dispersed material displacement decreases, and the trajectory changes to a pulse-forward one. The trajectory length of the dispersed material motion increases, the time of its contact with the drying agent is extended.

As the diameter of the perforation holes decreases, the effect of the drying agent's ascending flow increases, in which the pulse component of the dispersed material motion trajectory decreases, and the forward increases. Thus, the trajectory length of the dispersed material motion increases, and the contact time with the drying agent is extended. It should be noted that with the further reduction of the perforation hole diameter, the action of the drying agent's ascending flow begins

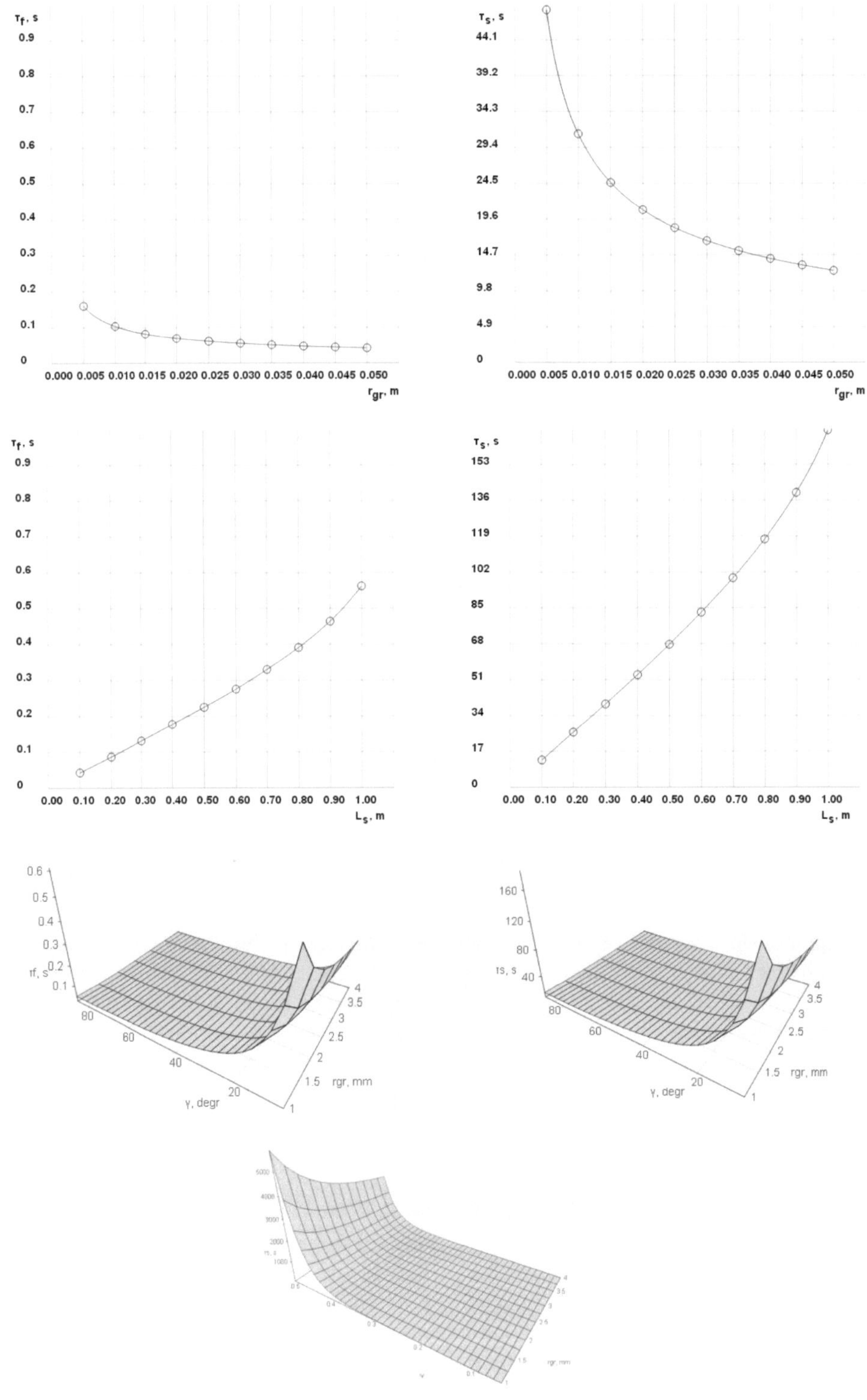

Figure 7.
Examples of calculation results.

significantly to outweigh the effect of the gravity rolling component. It leads to the formation of the second transitional mode and the ablation mode in the shelf dryer's operation.

3.2 Heat-mass transfer

The calculated values of the mass transfer coefficient β from Eq. (16) depending on the velocity of the drying agent's motion are demonstrated in **Figure 8**.

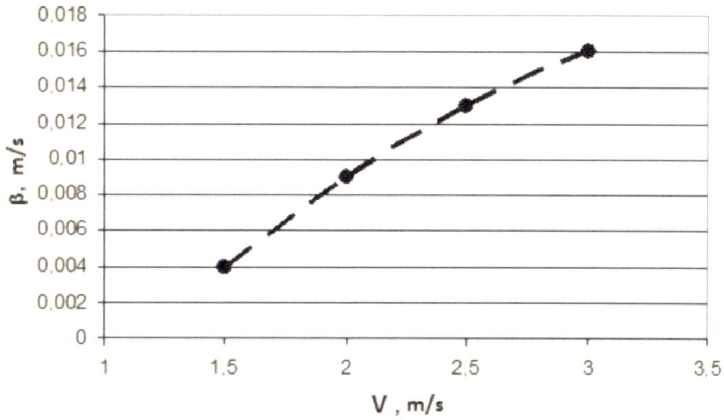

Figure 8.
Dependence of the mass transfer coefficient on the drying agent's motion velocity.

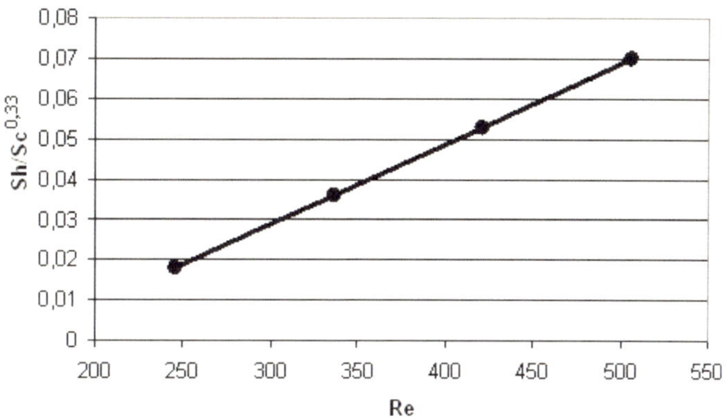

Figure 9.
Graphical dependency $Sh/Sc^{0.33} = f(Re)$ to define the coefficient A_1 and equation stage m.

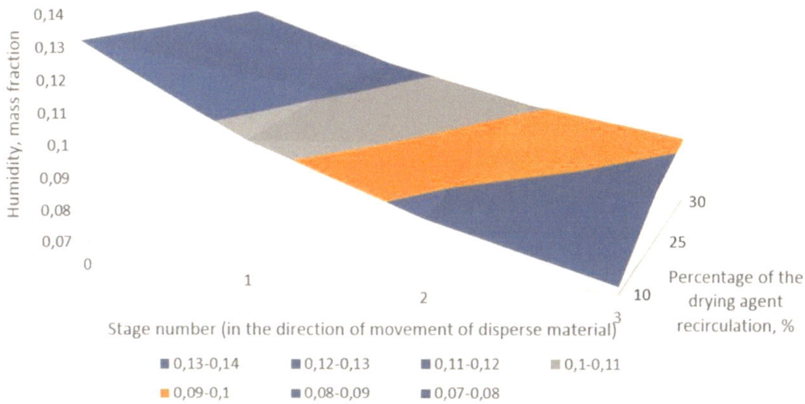

Figure 10.
Influence of the drying agent recirculation method on the change of the moisture content in the disperse material.

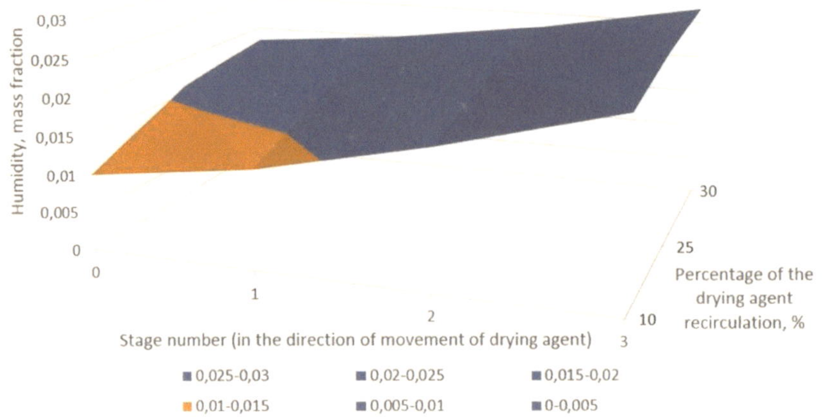

Figure 11.
Influence of the drying agent recirculation method on the change of the moisture content in the drying agent.

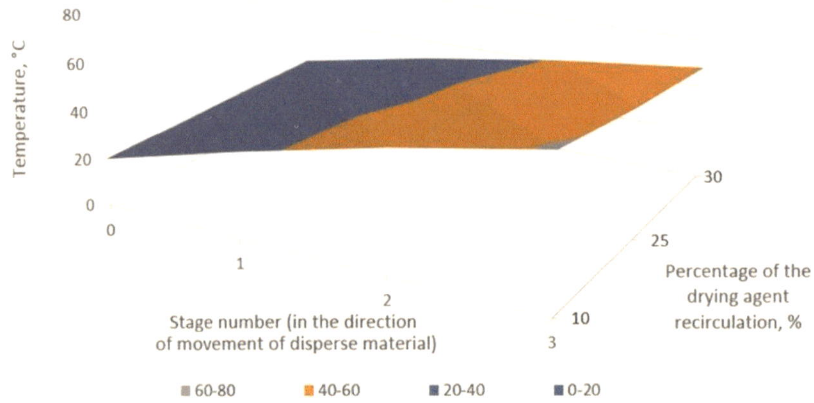

Figure 12.
Influence of the drying agent recirculation method on the temperature change of the disperse material.

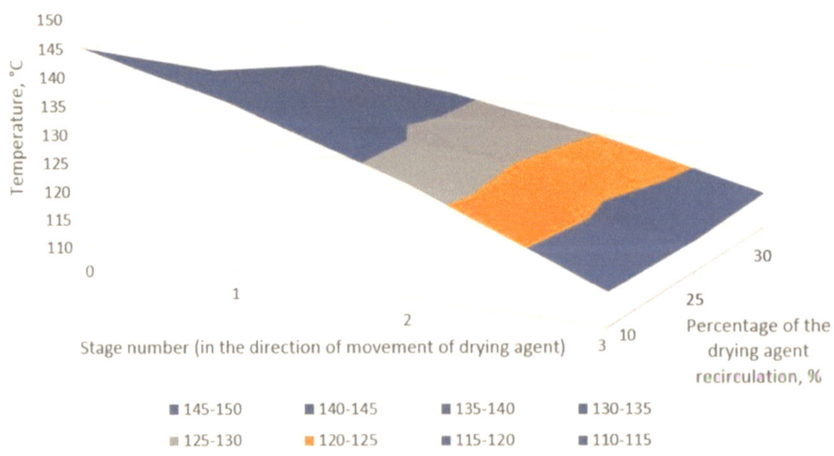

Figure 13.
An influence of the drying agent recirculation method on the temperature change of the drying agent.

The graphical dependency from Eq. (17) shows (**Figure 9**) that coefficient $A_1 = 0.008$, $m = 0.47$. The coefficient n is 0.33 for the situation when the drying agent's parameters were slightly changed during the experiment [8].

Taking into account the obtained values of the coefficient A_1 and equation stage m, the criterial value (18) will be as follows:

$$Sh = 0.008 \cdot Sc^{0.33} \cdot \mathrm{Re}^{0.47}. \tag{19}$$

3.3 Kinetics and drying efficiency

The organization of the drying agent's motion may have a considerable influence on the quality indicators of the dried material and the properties of the drying

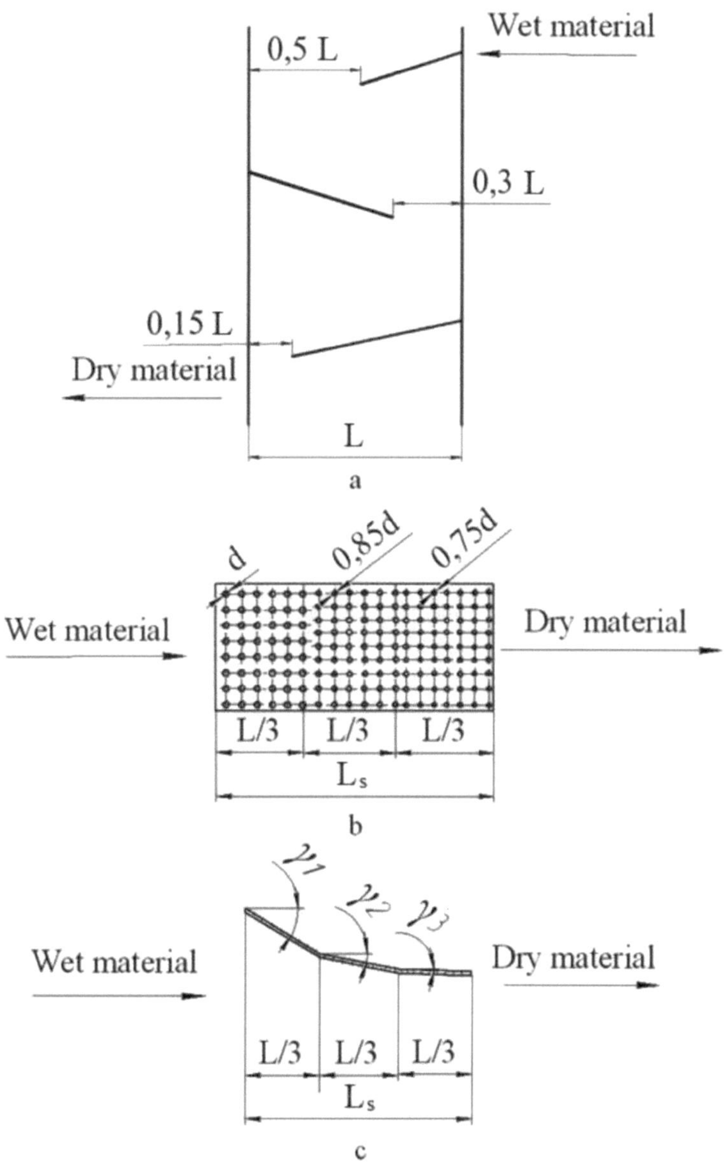

Figure 14.
Constructions of shelves in the multistage gravitational shelf dryer [13–15]: (a) shelf with a different gap on the height of dryer, (b) sectioned shelf with variable perforation of sections, and (c) partitioned sections shelf with constant perforation and variable angle of inclination.

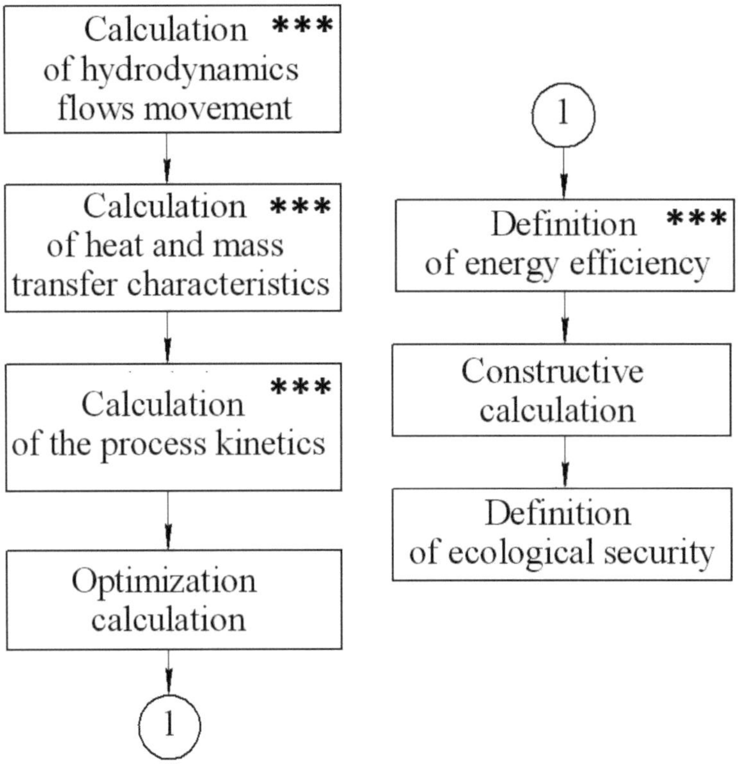

Figure 15.
*Block scheme of the algorithm to calculate the multistage gravitational shelf dryer (symbol *** shows the blocks which are described in this work).*

agent. That has evolved several studies, the results of which are presented in **Figures 10–13**. Their analysis enables us to select the method to organize the drying agent's motion, which consumes the least energy and ensures the necessary complete removal of moisture from the disperse material.

The analysis of the figures shows that the features of the dispersed material and the drying agent are changed according to one law; each of the technological indicators in the drying agent differently influences the intensity of the increase or decrease of parameters. The figures show that there is no function extremum on the graphical dependencies, which is explained by the regularities of the convective drying kinetics—the parameters' change of the contacting flows in each of the periods occurs monotonically with different intensity on separate sites depending on the dehydration conditions.

Different constructions of the shelves (**Figure 14**) enable us to control the residence time of the dispersed phase in the dryer's workspace.

Block scheme of the algorithm to calculate the multistage gravitational shelf dryer is represented in **Figure 15**.

4. Experimental research and practical implementation

During the optimization calculations, the necessity to obtain certain empiric dependencies has been revealed. They would identify some quantities, especially important for the shelf dryer's design development.

The experimental investigations were carried out on a shelf dryer model, the design parameters of which corresponded to the picture in **Figure 1**. Experiments to study the properties of the two-phase flow hydrodynamics in a shelf device were carried out at gas flow velocity of 1–5 m/s, specific capacity on the source material 6–10 kg/(m^2 s).

At low gas flow velocities of 0.5–1 m/s, particles of the material move along the surface of the sloping shelf at a velocity of 0.2–0.3 m/s in the form of a rapidly “skipping” layer. The particles of the material are braked at the wall of the device in the discharge space and are accumulated on the surface of the wall (**Figure 16a**) after moving over the surface of the shelf. This accumulated layer is blown by a gas jet, which is formed by a discharge gap between the lower end of the shelf and the wall of the device. As the gas flow velocity increases to 2.5–4 m/s, the porosity of the layer decreases to 0.75–0.8, and the concentration of particles in the layer increases to 40–50 kg/m^2 s.

Small particles, in which the inertia force during their discharge from the surface of the sloping shelf is insufficient to overcome the kinetic energy of the gas jet, are picked up by the jet and move along a curved path to the upper part of the device—the separation zone. On the photo (**Figure 16a**), it is seen by the distinct tracks of the trajectory. Large particles, overcoming the aerodynamic drag of the gas jet, fall out of the layer through the discharge space down. The described hydrodynamic regime is called “gravitationally falling layer.” This mode is implemented on the shelves, installed with a width of the discharge gap (0.3–0.5) L and a free area of 5–10%. Therefore, the maximum efficiency of small fraction ablation by the gas flow is achieved (**Figure 17**). The velocity of the complete ablation of the small fraction without large particles in it—the second critical velocity—is as follows:

$$Re_{2cr} = 0.1Ar^{0.7}, \text{ with } Ar \leq 62,000, \tag{20}$$

$$V_{cr} = \frac{\mathrm{Re}_{2cr} \cdot \nu}{d_{gre}} \tag{21}$$

where Ar is the Archimedes criterion, $Ar = \frac{d_{gre}^3 \cdot (\rho_{gr} - \rho_g) \cdot g}{\nu^2 \cdot \rho_g}$.

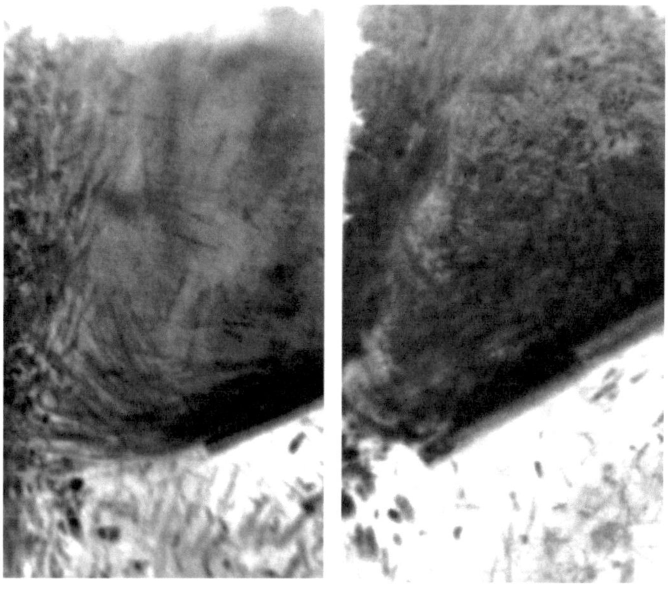

Figure 16.
The photo of the hydrodynamic modes of the shelf dryer: (a) “gravitationally falling layer” regime and (b) “weighted layer” regime.

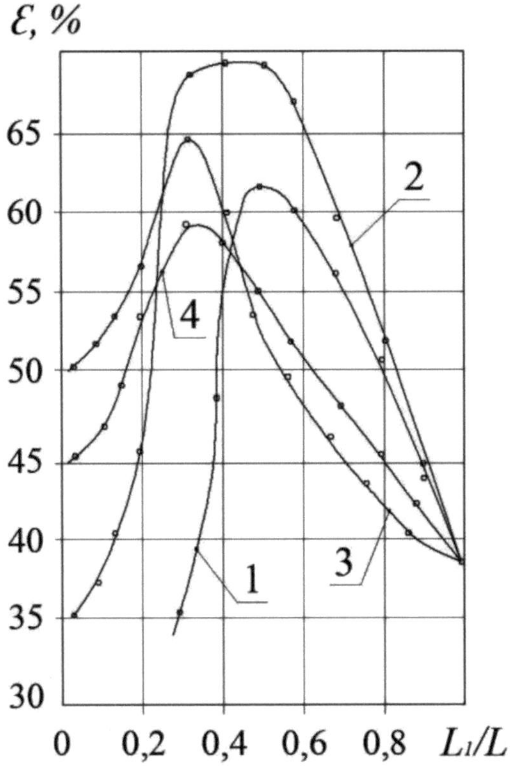

Figure 17.
The influence of the design (constructive) parameters of the shelf on the extraction efficiency of the fraction less than 1 mm. The free area of the shelf: 1–4, respectively, 0, 5, 15, and 30%. The tilt angle of the shelf is 30°. Material is a polydisperse mixture of the granular superphosphate.

Reducing the width of the discharge space up to (0.15–0.2) L and increasing the free area of the shelf up to 15%, owing to the growing kinetic energy of the gas jet, the continuously circulating vortex layer of particles above the sloping shelf surface is formed (**Figure 16b**). Therefore, the particles of the material are moving on the surface of the sloping shelf in the form of the tightened (compressed) layer at the velocity of 0.05–0.15 m/s, and in the area above, the discharge space—in the form of weighted, intensively circulating layer. The porosity of this layer is 0.65–0.7 (coincides with fluidized systems' porosity), and concentration of the particles in the workspace of the device is 160–280 kg/m^3. The described hydrodynamic regime is called the "weighted layer." The velocity at which the "weighted layer" mode is implemented—the critical velocity when the weighting is started is calculated as follows:

$$R_{wl} = Re_w \cdot k \cdot (L1/L), k = 1.19 \cdot \, lg \left(100 \cdot f_a\right) + 0.005, \tag{22}$$

$$V_{wl} = \frac{\mathrm{Re}_{wl} \cdot \nu}{d_{gre}}, \tag{23}$$

where Re_w is the Reynolds criterion for particle staying in the gas flow, $\mathrm{Re}_w = \frac{V_w \cdot d_{gre}}{\nu}$; V_w is the velocity of the medium-size particle hovering; f_a is the free area of the shelf (%).

The effect of the gas flow velocity on the intensity of the interphase heat transfer process is represented by the dependence of the Nusselt criterion on the Reynolds

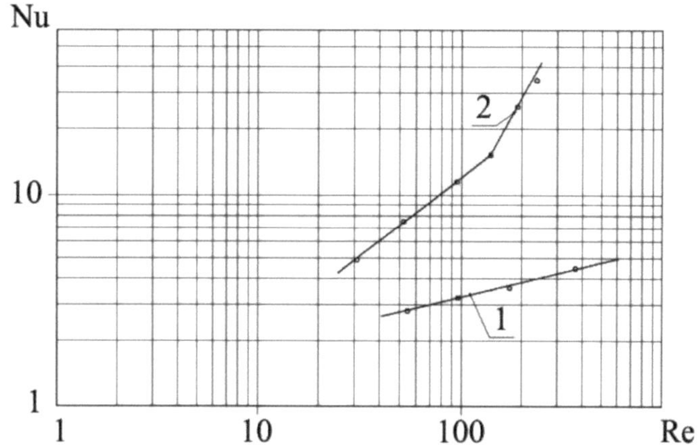

Figure 18.
Influence of the gas flow velocity on the interphase heat transfer intensity: (1) "gravitationally falling layer" mode, shelf parameters: L_1/L = 0.5; f_a = 5%; (2) "weighted layer" mode, shelf parameters: L_1/L = 0.15; f_a = 15%.

criterion—Nu = f (Re) (**Figure 18**). These dependencies are described with the following criteria equations:

$$\text{“gravitationally falling layer mode”} Nu = 1.5 \cdot Re^{0.21}, \tag{24}$$

$$\text{“weighted layer”mode } Nu = 0.38 \cdot Re^{0.73}. \tag{25}$$

Equations (24) and (25) are valid for $0 \leq \text{Re} \leq 500$.

The above dependencies in the "weighted layer" mode show that the values of the Nusselt criterion are significantly higher than values, which are peculiar for the "gravitationally falling layer" mode. The sufficiently high intensity of heat and mass transfer processes in the "weighted layer" mode is explained by the fact that in this mode the gas jet entering the weighted layer through the discharge gap at sufficiently high velocity has the greatest intensifying effect. Measures of the single-phase flow velocity, carried out by a thermal anemometer in the intersection above the shelf, showed that at gas jet velocities in the discharge space of 6–12 m/s, local heat transfer coefficients at the site of material particles contact with the gas jet are 400–500 W/(m^2 K). These values are peculiar for the intensive heat transfer conditions in the core of the spouting layer and exceed the average heat transfer coefficients for fluidized beds (100–400 W/(m^2 K)) and the pneumatic transportation mode (100–200 W/(m^2 K)).

The rapid evaporation of moisture in the zone above the discharge space leads to some temperature drop of the hot gas before it contacts with the main layer of particles, weighted above the shelf. Experimental studies show that the heating temperature of particles in the zone of contact with the gas jet entering the discharge gap is 1.5–2.0 times higher than that in a weighted layer on the surface of the shelf. It uses a drying agent with higher inlet temperature (1.5–1.8 times higher than melting temperature) than it is acceptable for dryers of the fluidized bed, without fear of the thermal damage of particles.

The investigated construction of the shelf dryer was tested when drying the fine- and coarse-crystalline potassium chloride, sodium pyrosulfite, iron, and nickel powders (**Table 1**).

The shelf dryer, where experimental tests were carried out, is a vertical rectangular-sectioned shaft, inside which the sloping perforated shelves are located in cascade on opposite sides (**Figure 1**). Wet material is fed by the batcher to the

Material	The velocity of the drying agent, m/s	Humidity, % wt			Moisture removal intensity, kg/(m^3 h)
		Source material	Undershooting	Ablation	
Fine-coarse potassium chloride:					
	1.32	6.1	0.35	0.24	421
Coarse-crystalline	1.5	6.1	0.1	0.06	462
	1.45	8.0	1.2	0.2	1025
	1.9	7.0	0.1	0.17	527
	2.1	8.0	0.11	0.1	1173
	2.3	7.0	0.14	0.1	528
Sodium pyrosulfite	2.8	5.0	0.5	0.1	250
	3.3	5.0	0.5	0.1	346
	3.5	6.0	0.6	0.2	258
Iron powder	2.0	10.0	0.34	0.3	1826
Nickel powder	2.3	11.3	0.4	0.3	600

Table 1.
Results of the drying of the granular materials in the shelf dryer.

upper shelf, is weighed above it, and is divided into small and large fractions. The upper shelf works in the hydrodynamic regime of the "gravitationally falling layer." In this mode, the dedusting process of materials, i.e., the removal of small particles from the initial mixture by the minimum interface, is effectively carried out. The minimum interface for shelf devices is 50–70 μm. The small particles are carried away by the drying gas agent into the separation space and then captured by a cyclone in which they are dried. Large particles fall down through the discharge space to the lower shelf.

When drying the materials, which are prone to the formation of lumps and strongly sticking to surfaces, the distance from the delivery point of the wet product into the device to the upper shelf has to be at least 0.3–0.5 m. The material, passing this distance, breaks up into small pieces and is partially dried. A hydrodynamic regime of the "weighted layer" is created on the lower shelf, in which, due to the intensive circulation and mixing of particles in the layer, the drying process is effectively carried out. The longer residence time of the particles in this layer also contributes to it.

The wet material, discharged from the upper shelf, enters the lower shelf from the top of the weighted layer, is drawn into the circulation, and is dried quickly. The share of the dried material falls through the discharge space into the hopper, in which a large fraction of the dried product is collected.

Thanks to the shelf contact elements, shown by the data of **Table 1**, the drying process takes place at the drying agent's moderate velocities (maximum 3.5 m/s) and at a large moisture intensity of the dryer's workspace which is up to 1000–1500 kg/(m^3 h). Due to the intensive contact between phases in the shelve devices, the drying process is carried out at high specific loads of up to 15–20 kg/(m^2 s), significantly exceeding the specific loads of 0.1–1.5 kg/(m^2 s) for fluidized bed devices. The specific consumption of the drying agent in the shelf dryers reaches

0.5–0.7 m^3/kg, and the hydraulic resistance is 1300–1500 Pa, respectively, against the values of 1.4–2.8 m^3/kg and 1800–2200 Pa for fluidized bed devices. The working path of the pneumatic pipe dryer, in which energy is expended to accelerate and to lift the drying material, has a hydraulic resistance of 1600–2000 Pa.

An additional advantage of shelf dryers is the simultaneous dedusting of the drying material. Fine-crystalline potassium chloride, containing 7–10% of the small fraction with a particle size of less than 100 μm in the initial mixture, after processing in a shelf device at a gas flow velocity of 1.4–1.5 m/s, had 1.2–5.5% of the small fraction in the final product (undershooting) and 60–80% of the small fraction in ablation. Coarse-crystalline potassium chloride, containing 4–10% of the small fraction with a particle size of less than 100 μm in the initial mixture, had 2–5% of the small fraction in the final product (undershooting) at a gas flow velocity of 1.3–1.4 m/s and 58–65% of the small fraction in the ablation. The extraction degree of the small fraction into the ablation was 70–90%. When the gas flow velocity exceeds 1.5 m/s in the final product, the small fraction is practically absent, and the content of the coarse fraction (more than 100 μm) in ablation is 3–5%.

The small fraction was completely extracted from the polydisperse mixture of granulated superphosphate containing up to 20% of the small fraction with particle sizes less than 1 mm, after processing in the shelf device at a gas flow velocity of 3.5–3.8 m/s, into the ablation. Therefore, the extraction degree of the small fraction into ablation was 80–85%.

In order to prevent the coarse fraction ablation by the gas flow and increasing the residence time of particles in the separation space for drying the ablative fractions, the upper section with a constant intersection was replaced with a conical free intersection [16] or with shelf contact elements [17].

Thus, the shelf dryers achieve the higher technological effect than typical constructions of the fluidized bed dryers and pneumatic pipe dryers, with less energy, capital costs, and sizes.

5. Conclusions

The convective shelf dryer construction with active aerodynamic processing modes is proposed. The developed engineering method for the shelf dryer calculation made it possible to define the constructive parameters of the device, ensuring the minimum required drying time of the wet material in the device to a predetermined humidity index. The demonstrated author's program Multistage Fluidizer® for computer implementation of the engineering calculation method made it possible to optimize the constructive and operating parameters of the drying process in the shelf device. It was shown that a shelf dryer should have, for example, three shelf contacts with various widths of the discharge space and various free areas of the shelves.

The author shows various hydrodynamic regimes to weigh particles of a material by a gas flow, depending on the constructive parameters of shelf contacts.

The effectiveness of the shelf device to carry out the drying and dedusting processes of granular and powder materials simultaneously was experimentally proven.

Acknowledgements

This research work has been supported by the Slovak Grant Agency VEGA Grant No. 1/0731/16 "Development of Modern Numerical and Experimental

Methods of Mechanical System Analysis," by Cultural and Educational Grant Agency of the Slovak Republic (KEGA) Project No. KEGA 002TnUAD-4/2019, and by the Ministry of Science and Education of Ukraine under the project "Small-scale energy-saving modules with the use of multifunctional devices with intensive hydrodynamics for the production, modification and encapsulation of granules," Project No. 0119U100834.

Conflict of interest

The authors declare that they have no competing interests.

Author details

Artem Artyukhov[1*], Nadiia Artyukhova[1], Ruslan Ostroha[1], Mykola Yukhymenko[1], Jozef Bocko[2] and Jan Krmela[3,4]

1 Sumy State University, Sumy, Ukraine

2 Technical University of Košice, Košice, Slovak Republic

3 Alexander Dubcek University of Trencin, Puchov, Slovak Republic

4 University of Pardubice, Pardubice, Czech Republic

*Address all correspondence to: a.artyukhov@pohnp.sumdu.edu.ua

References

[1] Delgado JMPQ, Barbosa de Lima AG, editors. Transport Phenomena and Drying of Solids and Particulate Materials. Switzerland: Springer International Publishing; 2014. 115p

[2] Kowalski SJ, editor. Drying of Porous Materials. Netherlands: Springer; 2007. 231p

[3] Sazhin BS, Sazhin VB. Scientific Principles of Drying Technology. USA: Begell House Publishers Inc.; 2007. 509p

[4] Kowalski SJ. Thermomechanics of Drying Processes. Germany: Springer-Verlag Berlin Heidelberg; 2003. 358p

[5] Delagado JMPQ, Barbosa de Lima AG, editors. Drying and Energy Technologies. Switzerland: Springer International Publishing; 2016. 228p

[6] Mujumdar AS, editor. Handbook of Industrial Drying. 4th ed. USA: CRC Press Taylor & Francis Group; 2014. 1348p

[7] Law CL, Azharul K, editors. Intermittent and Nonstationary Drying Technologies: Principles and Applications. 1st ed. USA: CRC Press Taylor & Francis Group; 2017. 244p

[8] Artyukhova NA, Shandyba AB, Artyukhov AE. Energy efficiency assessment of multi-stage convective drying of concentrates and mineral raw materials. Naukovyi Visnyk Natsionalnoho Hirnychoho Universytetu. 2014;**1**:92-98

[9] Artyukhov AE, Sklabinskyi VI. Experimental and industrial implementation of porous ammonium nitrate producing process in vortex granulators. Naukovyi Visnyk Natsionalnoho Hirnychoho Universytetu. 2013;**6**:42-48

[10] Kudra T, Mujumdar AS. Advanced Drying Technologies. 2nd ed. USA: CRC Press; 2009. 438p

[11] Artyukhova NA. Multistage finish drying of the N_4HNO_3 porous granules as a factor for nanoporous structure quality improvement. Journal of Nano- and Electronic Physics. 2018;**10**(3): 03030-1-03030-5

[12] Artyukhov AE, Artyukhova NO, Obodyak VK, Horishnyak AO. Certificate of authorship No. 79141 (Ukraine). Computer program Multistage Fluidizer©

[13] Artyukhova NO, Yukhymenko MP, Artyukhov AE, Shandyba AB. Patent No. 74070, (Ukraine). Device for drying of disperse materials

[14] Artyukhova NO, Yukhymenko MP, Artyukhov AE, Shandyba AB. Patent No. 81720, (Ukraine). Device for drying of disperse materials

[15] Artyukhov AE, Artyukhova NO, Shandyba AB. Patent No. 92423, (Ukraine). Device for drying of disperse materials

[16] Yukhymenko M, Ostroha R, Litvinenko A, Bocko J. Estimation of gas flow dustiness in the main pipelines of booster compressor stations. IOP Conference Series: Materials Science and Engineering. 2017;**233**:012026

[17] Lytvynenko A, Yukhymenko M, Pavlenko I, Pitel J, Mizakova J, Lytvynenko O, et al. Ensuring the reliability of pneumatic classification process for granular material in a rhomb-shaped apparatus. Applied Sciences. 2019;**9**:1604

Chapter 4

Mathematical Modeling and Simulation of Rapeseed Drying on Concurrent-Flow Dryer

Le Anh Duc and Keum Dong Hyuk

Abstract

Mathematical modeling for rapeseed drying on concurrent-flow dryer was built based on energy and mass transfer balances. The fourth-order Runge–Kutta method was used for solving four ordinary differential equations. A computer simulation program for circulating concurrent-flow rapeseed dryer was developed using these models. A pilot-scale concurrent-flow dryer was used to verify the fitness of simulation program. Two drying experiments were conducted. The output parameters of the simulation program were compared and analyzed with experiment data. The RMSE of simulated moisture contents ranged from 0.334 to 0.506%w.b. with the coefficient of determinations ranged from 0.994 to 0.997. The RMSE of simulated rapeseed temperatures during drying process ranged from 1.15 to 1.77°C with the R^2 ranging from 0.904 to 0.925. The experimental drying rates were 2.38 and 2.80% w.b./h. In comparison with simulated values, the difference between simulated value and measured value of drying rate were 5.04 and 5.08%; drying time were 7.14 and 0.47%; and germination ratio were 1.87 and 0.47%. The simulated fuel energy consumption for drying were 4.62 and 8.57% lower than the experimental values. The analytic results showed that the simulation results have good fitness with experimental data.

Keywords: concurrent-flow dryer, mathematical modeling, simulation, drying rate, grain temperature, moisture content

1. Introduction

There are many different grain dryer designs on the market. Basically, the classical configurations of the moving bed dryers fall into four categories according to the relative directions of seed and air flows: the mixed flow dryer, the cross flow dryer, the concurrent-flow dryer, and the countercurrent flow dryer. The configuration with parallel concurrent-flow displays some advantages, such as obtaining of more homogeneous products as well as a better energy usage.

Concurrent-flow drying is a relatively new grain drying technology. In a concurrent-flow dryer, both the grain and drying air are moving in the same direction. This type of dryer has the advantage of using very high drying air temperatures without affecting grain quality and does not suffer the variation in grain moisture contents. Besides, energy efficiency of this type dryer is high.

The concurrent-flow drying principle was introduced the first time in 1955 by Öholm. Thompson et al. [1] developed simulation models of concurrent-flow dryer for corn drying. Now most of the drying simulation model was implemented based on the model presented by Thompson et al. that was developed for simulating high temperature corn drying. Felipe and Barrozo [2] studied the simultaneous heat and mass transfer between air and soybean seeds in a concurrent moving bed dryer, based on the application of a two-phase model to the drying process. Keum et al. [3] studied on circulating concurrent flow for rice drying with the drying temperature from 98 to 126°C, and air flow rate from 28.5 to 57.1 cmm/m^2. The study results showed that drying rate ranged from 1.09 to 2.2% d.b./h, and energy consumption ranged from 6224 to 6992 kJ/kg-water.

From these advantages, the use of concurrent-flow drying principle for drying of rapeseed has been recommended owing to:

- Energy saving: energy efficiency of this type dryer is about 30–40% better than a cross flow type dryer without heat recovery.
- It is possible to use a high drying temperature (up to 130°C) without increasing the grain temperature excessively because grain are exposed to drying air in a short time, leads to high drying rate (1.5–2% w.b./h).
- The drying air flow is parallel to the grain flow leads to more homogeneous moisture content and temperature distributions because all grains are exposed to the same temperatures, therefore guaranteeing the quality of the dried grains.

Mathematical modeling and computer simulation can be used to predict the moisture and temperature of the products during the drying process and the energy consumption and drying capacity of the different drying system. Many different models have been proposed to describe the drying process in the basic types of convective grain dryer. However, no previous research was found in literature that study on dynamic simulation of concurrent flow for rapeseed.

2. Concurrent-flow drying model

2.1 Selection of drying principle

The drying process of rapeseed in concurrent-flow dryer was described in **Figure 1**.

The hot air and rapeseed are moving the same direction in the drying chamber. At the end of cycle drying, hot air is exhausted to ambience by suction fan; rapeseed is continued to be circulated on the drying chamber for the next cycle drying until it reaches the desired final moisture content.

2.2 Mathematical model

The mathematical model used in the study consists of a set of four partial differential equations in four independent variables: air humidity, air temperature, grain temperature, and grain moisture content.

Based on the theory of energy and mass transfer, a concurrent-flow rapeseed drying model was developed. By using this mathematical models, performance of concurrent-flow rapeseed dryer can be predicted; the temperature and moisture

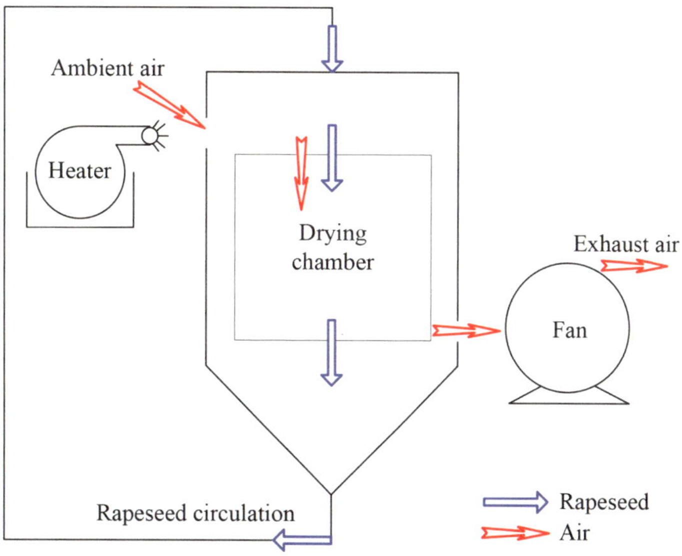

Figure 1.
Block diagram of rapeseed concurrent-flow dryer.

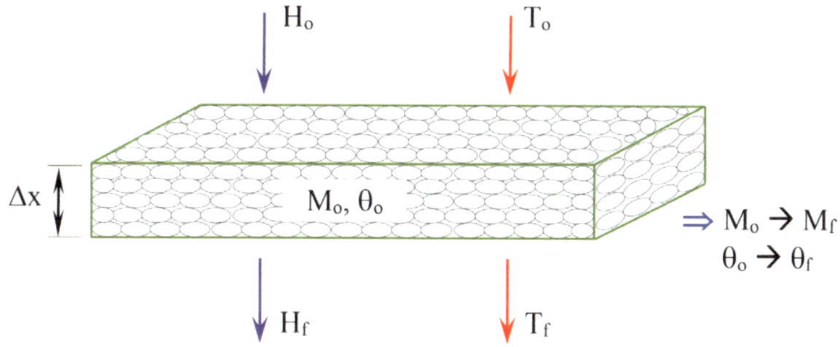

Figure 2.
Concurrent-flow drying model.

content of rapeseed and the temperature and relative humidity of drying air were predicted.

The thin layer drying models is used in simulation of deep-bed dryer, in which the average changes in moisture content and temperature on a thin layer of grain are calculated over a discrete time interval Δt. In the simulation, to solve mathematical models of deep-bed drying process, the depth bed was divided into nth thin layers with a thickness of Δx each, and dynamic heat and mass balances were set up in each section, and then the model consisted in a set of partial differential equations (**Figure 2**). Drying is achieved by continuously passing hot air through the static grain bed in one direction, from the top to the bottom section. From the first thin layer at the top section, the air evaporates moisture from the grain and carries it to the next thin layer. As the drying air absorbs moisture, its temperature is decreased, and its ability to pick up more moisture (drying potential) decreases. A deep bed consisting of a number of thin layers is, therefore, simulated by calculating the air and moisture changes as the drying air passes from one thin layer of grain to the next layer. Each layer dries to equilibrium conditions for short time intervals, and the exhaust air from one layer is used as the input drying air to the next. The drying procedure continues for a number of drying time intervals until the desired final moisture content of the material is achieved.

Some of these assumptions are made to simplify the mathematical model. So, for developing these mathematical models, the following assumptions were made:

- Operation is in the steady state.
- Grain shrinkage is negligible during the drying process.
- No temperature gradients exist within each grain particle.
- Particle-to-particle conduction is negligible.
- Initial moisture content of grain is uniform.
- Airflow and grain flow are plug-type and constant.
- The dryer walls are adiabatic and heat losses are negligible.
- The heat capacities of moist air and of grain are constant during the short time periods.
- The solids flow rate is uniform.

Energy balances and mass balances are written on a differential volume (S·Δx) located at an arbitrary location in the grain bed. There are four unknowns in this problem:

T(x,t): the air temperature.

θ(x,t): the grain temperature.

H(x,t): the humidity ratio.

M(x,t): the grain moisture content.

Therefore, four equations for material and energy balances must be made in order to calculate of T, θ, H, and M. Resulting from the balances are four Eq. (4):

1. For the enthalpy of the air (air enthalpy balance):

The change in sensible heat of air that results due to heat transfer by convection in time. The air enthalpy balance over the differential volume:

energy out = energy in – energy transferred by convection

$$\frac{dT}{dx} = \frac{-h_c a}{G_a c_a + G_a c_v H}(T - \theta) \quad (1)$$

2. For the enthalpy of the grain (grain enthalpy balance):

The enthalpy from the air to the grains due to convection heat transfer over the control volume is equal to the required heat to evaporate water inside the grains and to heat water vapor extracted from the grains and the rate of accumulated heat inside the grains. The grain enthalpy balance over the differential volume:

energy transferred = change in internal grain energy – energy for evaporation

$$\frac{d\theta}{dx} = \frac{h_c a}{G_p (c_p + M c_w)}(T - \theta) - \frac{h_{fg} + c_v (T - \theta)}{G_p (c_p + M c_w)} G_a \frac{dH}{dx} \quad (2)$$

3. For the humidity of the air (humidity balance):

The amount of water in and out the differential volume is equal to the rate of change of moisture content in the grains.

$$\text{moisture transferred} = \text{moisture in–moisture out}$$

$$\frac{dH}{dx} = -\frac{G_p}{G_a}\frac{dM}{dx} \tag{3}$$

4. For the moisture content of the grain (thin layer drying equation):

$$\frac{dM}{dx} = \text{an appropriate thin layer drying equation}$$

$$\frac{dM}{dx} = \frac{M_o - M_e}{v_p}\left(-Pt^{Q-1}Q\right)\exp\left(-Pt^{Q}\right) \tag{4}$$

where:
a: specific surface area of grain (m^2/m^3).
c_a: specific heat of dry air (kJ/kg·K).
c_p: specific heat of dry grain (kJ/kg·K).
c_v: specific heat of water vapor (kJ/kg·K).
c_w: specific heat of water in grain (kJ/kg·K).
G_a: air flow rate ($kg/h{\cdot}m^2$).
G_p: grain flow rate ($kg/h{\cdot}m^2$).
h_{fg}: vaporization latent heat of water within grain (kJ/kg).
H: enthalpy of dry air (kJ/kg).
h_c: convection heat transfer coefficient ($kJ/h{\cdot}m^2{\cdot}{}^oC$).
v_p: grain velocity (m/h).

Eqs. (1) and (2) represent the respective energy balances. Eqs. (3) and (4) result from mass balances applied for both fluid and solid phases of a concurrent-flow dryer.

The equation for the moisture content is obtained from the empirical thin layer equation for rapeseed. The four differential equations from (1)–(4) constitute the concurrent-flow drying model. A computer simulation program was developed using these models.

To solve these differential equations, the initial and boundary condition of grain and the drying air must be known and furnished to the simulation program as input data [4]. In this category fall:

- The initial or inlet temperature and moisture content of grain
- The initial or inlet temperature and absolute humidity of the drying air

The initial conditions of the air humidity, grain moisture, and both air and grains temperatures were assumed constant at the dryer inlet, resulting in the following model boundary conditions (x = 0) for concurrent-flow dryer which are:

$$T_{x=0} = T_{in} \tag{5}$$

$$\theta_{x=0} = \theta_o \tag{6}$$

$$H_{x=0} = H_{in} \tag{7}$$

$$M_{x=0} = M_o \quad (8)$$

where:
T_{in}: inlet air temperature (°C).
θ_o: initial grain temperature (°C).
H_{in}: inlet humidity ratio of drying air (kg/kg).
M_o: initial moisture content of grain (dec., d.b.)

2.3 Related equations

The related equations used for simulation such as specific surface area, latent heat, convection heat transfer coefficient, equilibrium moisture content, and thin layer drying equation of rapeseed were taken from specific studies.

2.3.1 Specific surface area of rapeseed

Specific surface area of rapeseed was determined in Eq. (9) [5]:

$$a = \frac{6(1-\varepsilon)}{d} \quad (9)$$

where:
ε: void fraction was calculated based on bulk density and true density of rapeseed, $\varepsilon = 0.389$.
d: diameter of rapeseed was determined in our previous study, $d = 2.21 \times 10^{-3}$ m [6]. Then, the specific surface area of rapeseed is $a = 1659\ m^2/m^3$.

2.3.2 Latent heat

Gallaher [7] established the following equation Eq. (10) to determine the dependence of the latent heat of vaporization of water from the product on its moisture content:

$$h_{fg} = h_{fgo} \bullet [1 + A \bullet \exp.(-B \bullet M)] \quad (10)$$

h_{fg}: latent heat of vaporization of water in grain (kJ/kg-water).
h_{fgo}: latent heat of vaporization of free water (kJ/kg-water).

$$h_{fgo} = (2502.2 - 2.386 \bullet T)$$

T: drying air temperature (°C).
M: rapeseed moisture content (decimal, d.b.)
A, B: coefficients.
The coefficients A and B were determined based on equilibrium moisture content of rapeseed (Modified Halsey equation) [8].
The values of the relative humidity can be replaced by equilibrium relative humidity (ERH) obtained from the equilibrium moisture content versus ERH relationships described in above section. Then:

$$P_v = P_s \bullet ERH \quad (11)$$

The relationship between vapor pressure and latent heats of two substances at the same temperature is as follows [5]:

$$\ln P_v = \frac{h_{fg}}{h_{fgo}} \ln P_s + C \tag{12}$$

where C is a constant of integration.

The ratio of heat of vaporization of water in grain to the heat of vaporization of saturated water on a logarithmic scale gives the ratio of latent heat h_{fg}/h_{fgo} at each level of moisture content as the slope of the straight line obtained from Eq. (12):

$$\frac{h_{fg}}{h_{fgo}} = 1 + A \cdot \exp(-B \cdot M) \tag{13}$$

Using MATLAB simulation program, the coefficients were found.

A = 0.3734.

B = 14.2442.

This result is fairly similar to the result of Cenkowski et al. [9]: A = 0.5; B = 14.5.

Convection heat transfer coefficient [5]:

$$h_c = 0.277 \cdot G_a \cdot \left(\frac{d \cdot G_a}{\mu_a}\right) \tag{14}$$

where:

h_c: convection heat transfer coefficient of grain bed (kJ/h·m^2·K).

d: geometric mean diameter of rapeseed (m).

μ_a: dynamic viscosity of air (kg/h·m).

$$\mu_a = 0.06175 + 0.000165 \cdot T \tag{15}$$

T: drying temperature (K).

Equilibrium moisture content [8]:

Equilibrium moisture content of rapeseed was determined using Modified Halsey equation:

$$M = [\exp(-4.9758 - 0.0132 \cdot T)]^{1/1.8755}(-\ln RH)^{-1/1.8755} \tag{16}$$

Thin layer drying [10]:

Thin layer drying equation of rapeseed was determined using model of Page:

$$\mathrm{MR} = \exp.(-P \cdot t^Q) \tag{17}$$

where the drying constants were determined:

$P = 0.02246 + 3.2428 \cdot RH + 0.0006308 \cdot T^2 - 2.01481 \cdot RH^2 - 0.06077 \cdot T \cdot RH$.

$Q = 0.60932 - 1.72018 \cdot RH^2 + 0.02529 \cdot T \cdot RH$.

2.4 Numerical solution

Eqs. (1)–(4) and the boundary conditions Eqs. (5)–(8) were solved using the Runge–Kutta methods.

In numerical analysis, the Runge–Kutta methods are an important family of iterative methods for the approximation of solutions of ordinary differential equations (ODEs). These techniques were developed around 1900 by the German mathematicians Runge C. and Kutta M.W. One member of the family of

Runge–Kutta methods that is so commonly used is the fourth-order Runge–Kutta method or also called as RK4, meaning that the error per step is on the order of h^5, while the total accumulated error has order h^4 [11].

Let an initial value problem be specified as follows:

$$y' = f(y, t), y(t_0) = y_0 \tag{18}$$

where $f(t,y)$ is a function of y and t and the second equation is an initial condition.

In order to calculate y_{n+1} with a known value of y_n, integrate Eq. (18) in the interval $t_n \geq t \geq t_{n+1}$ to yield.

$$y_{n+1} = y_n + \int_{t_n}^{t_{n+1}} f(y, t)dt \tag{19}$$

The RK4 method is derived by applying a numerical integration method to the right side of Eq. (19). Then, the general form of the RK4 method for this problem is given by the following equations:

$$y_{n+1} = y_n + \frac{1}{6}(k_1 + 2k_2 + 2k_3 + k_4) \tag{20}$$

$$t_{n+1} = t_n + h \tag{21}$$

for n = 0, 1, 2, 3, ...

$$k_1 = hf(y_n, t_n) \tag{22}$$

$$k_2 = hf\left(y_n + \frac{k_1}{2}, t_n + \frac{h}{2}\right) \tag{23}$$

$$k_3 = hf\left(y_n + \frac{k_2}{2}, t_n + \frac{h}{2}\right) \tag{24}$$

$$k_4 = hf(y_n + k_3, t_n + h) \tag{25}$$

h: size of the interval

Thus, the next value (y_{n+1}) is determined by the present value (y_n). The slope is a weighted average of slopes:

k_1 is the slope at the beginning of the interval.

k_2 is the slope at the midpoint of the interval, using slope k_1 to determine the value of y at the point (t_n + h/2) using Euler's method.

k_3 is again the slope at the midpoint, but now using the slope k_2 to determine the y-value.

k_4 is the slope at the end of the interval, with its y-value determined using k_3.

In averaging the four slopes, greater weight is given to the slopes at the midpoint:

$$slope = \frac{1}{6}(k_1 + 2k_2 + 2k_3 + k_4) \tag{26}$$

To solve mathematical models of drying process, the depth bed model was divided into 10 thin layers, and the dynamic heat and mass balances were set up in each section and calculated over a discrete time interval Δt = 0.01 h.

3. Simulation program

3.1 Main program

The numerical solution for four ordinary differential equations was obtained by using MATLAB code programs based on fourth-order Runge–Kutta methods. MATLAB is an interactive program and technical computing environment with numeric computation and data visualization. It provides integrated numerical analysis, matrix computation, signal processing, and graphics in an easy-to-use environment where problems and solutions are easily expressed without complicated programming. MATLAB-based software, entitled RCDSim-LAD (Rapeseed Concurrent-flow Drying Simulation, version LAD), was built for drying simulation.

The concurrent-flow dryer simulation model was programmed with the sequence:

- Input data
- Initialize arrays
- Evaluate constants
- Solve four ordinary differential equations
- Output when appropriate

The program of concurrent model terminates in one of two ways:

- When the moisture content within the dryer reaches a specified level
- When condensation or absorption is detected

However, condensation or absorption is not simulated in the concurrent model since it does not occur in a properly designed dryer. Equations used by more than one model (e.g., psychrometric equations) are programmed as separate subroutines of function subprogram. A computer simulation program was built using these models. This program was used in predicting the performance and temperature profile within the grain bed.

Input data	Output data
Initial grain condition:	Drying time (h)
• Initial grain moisture content (dec, w.b.)	Number of pass
• Initial grain temperature (°C)	Final moisture content (% w.b.)
• Desired final moisture content (dec,w.b.)	Drying rate (% w.b./h)
Dryer specification:	Water removal rate (kg/m^2)
• Capacity (kg)	Fan static pressure (Pa)
• Drying air flow rate (cmm/m^2)	Fan power (kW/m^2)
• Grain flow velocity (m/h)	Fan energy (kJ/kg-water)
Drying and ambient air condition:	Fuel energy (kJ/kg-water)
• Drying air temperature (°C)	Total energy consumption (kJ/kg-water)
• Ambient air temperature (°C)	
• Ambient air relative humidity (dec.)	

Table 1.
Input and output data in the simulation program.

Table 1 listed input data and output data which was simulated by a simulation program. The flow chart of the simulation program is shown in **Figure 3**. For the convenience, the interface graphical user interface (GUI) was built in **Figure 4**.

3.2 Energy consumption

3.2.1 Fuel energy consumption

$$EnFuel = \frac{G_a \cdot (c_a + c_v \cdot h) \cdot (T - T_{amb}) \cdot t}{0.85 \cdot (M_o - M_f) \cdot \rho_g \cdot x} \quad (27)$$

where:
EnFuel: fuel energy consumption (kJ/kg-water).
G_a: air flow rate (kg/h·m^2).
c_a: specific heat of dry air (kJ/kg·K).
c_v: specific heat of water vapor (kJ/kg·K).
h: absolute humidity (kg-water/kg-dry air).
T: drying air temperature (°C).
T_{amb}: ambient air temperature (°C).
t: drying time (h).
M_o: initial moisture content (decimal, d.b.)
M_f: final moisture content (decimal, d.b.)
ρ_g: dry grain bulk density (kg/m^3).
x: grain layer thickness (m).

3.2.2 Fan power

$$PFan = \frac{\Delta P \cdot g_a \cdot A}{60 \cdot e_f \cdot 1000} \quad (28)$$

where:
PFan: fan power (kW).
ΔP: pressure drop (Pa).
A: cross-section area of grain bed (m^2).
g_a: air flow rate (m^3/min·m^2).
e_f: efficiency of the fan and motor (usually a value of about 0.5 is used).

3.2.3 Fan energy consumption

$$EnFan = \frac{PFan \cdot t \cdot 3600}{(M_o - M_f) \cdot \rho_g \cdot x} \quad (29)$$

where:
EnFan: fan energy consumption (kJ/kg-water).

3.2.4 Total energy consumption

$$\text{EnTot} = \text{EnFuel} + \text{EnFan} \quad (30)$$

where:
EnTot: total energy consumption (kJ/kg-water).

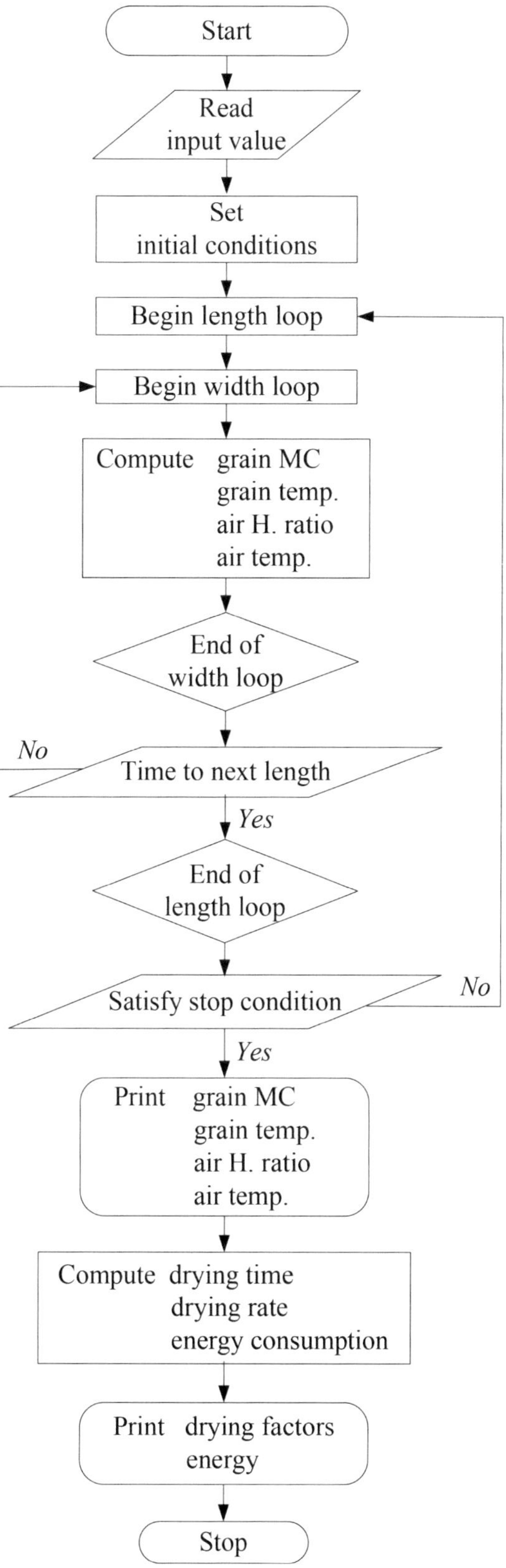

Figure 3.
Flow chart of the simulation program.

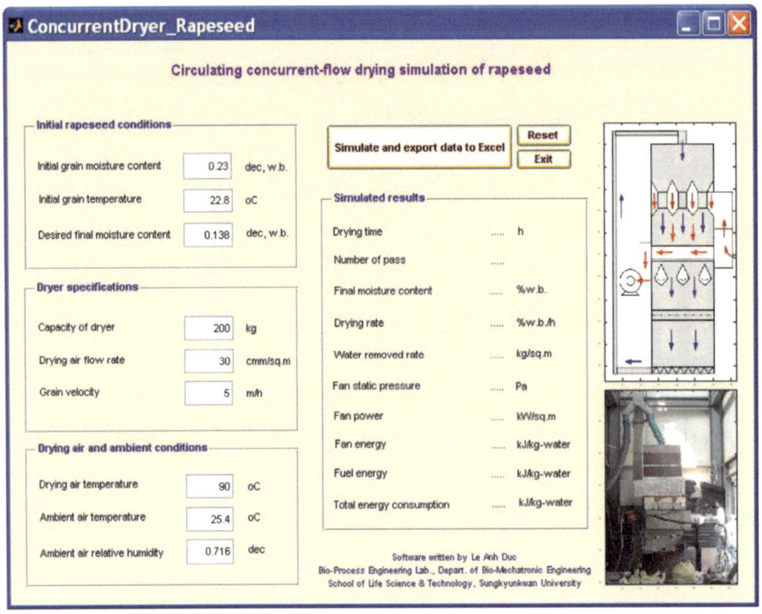

Figure 4.
Interface of the simulation program RCDSim-LAD.

3.3 Psychrometric properties

3.3.1 Saturated vapor pressure

$$P_s = r\frac{a + b \cdot T_k + c \cdot T_k^2 + d \cdot T_k^3 + e \cdot T_k^4}{f \cdot T_k - g \cdot T_k^2} \tag{31}$$

where:
P_s: saturated vapor pressure (Pa).
T_k: absolute temperature (K).
r = 22105649.25; a = −27405.526; b = 97.5413; c = −0.146244;
d = 0.12558×10^{-3}; e = -0.48502×10^{-7}; f = 4.34903; g = 0.39381×10^{-2}.

3.3.2 Absolute humidity

$$h = 0.6219 \cdot \frac{RH \cdot P_s}{P_{atm} - RH \cdot P_s} \tag{32}$$

h: absolute humidity (kg-water/kg-dry air).
RH: relative humidity (dec).
P_{atm}: atmospheric pressure, P_{atm} = 101,325 (Pa).

3.3.3 Specific volume

$$vs = 287 \cdot T_k \cdot \frac{0.6219 + h}{0.6219 \cdot P_{atm}} \tag{33}$$

vs.: specific volume (m^3/kg).

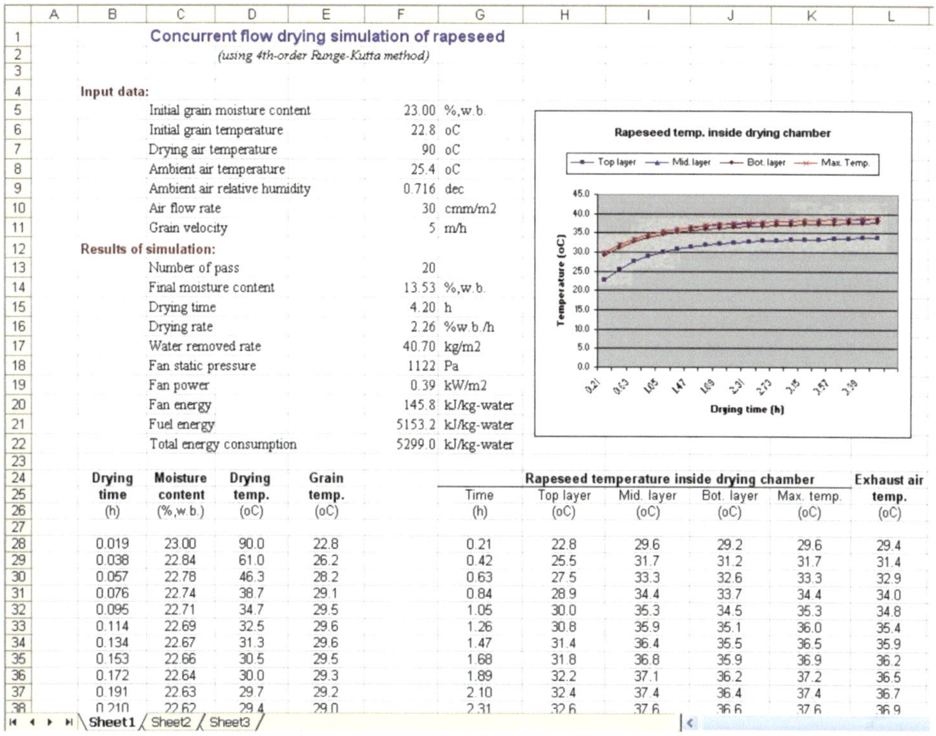

Figure 5.
Simulated data with excel interface.

Using MATLAB 7.3.0., Simulated results were exported to Excel (**Figure 5**) by clicking the button "Simulate and export data to Excel" on the interface of the simulation program (**Figure 4**).

In excel interface, besides the output data in the simulation program were listed in **Table 1**, the drying air temperature, rapeseed temperature, and moisture content of rapeseed in the drying process versus drying time are also displayed and automatically represented by graphs, in which the rapeseed temperature at the top layer, middle layer, bottom layer, and the maximum rapeseed temperature in the drying chamber versus drying time are also displayed.

4. Model validation

4.1 Pilot-scale concurrent-flow dryer

In order to verify a fitness of simulation program, a pilot-scale concurrent-flow rapeseed dryer with capacity of 200 kg/batch was designed and manufactured. The dimension of the dryer is shown in **Figure 6**, and the structural principle of the dryer is shown in **Figure 7**. The pilot-scale rapeseed dryer includes a drying tower with grain inlet section, plenum section, and drying section; burner; drying fan; variable speed discharge augers; and bucket elevator for circulating the grains.

The hot air is supplied by a kerosene jet burner; after going through the mixed chamber, the drying air will enter the dryer through plenum section. Rapeseed and drying air are moving the same direction until drying the air out by force of suction

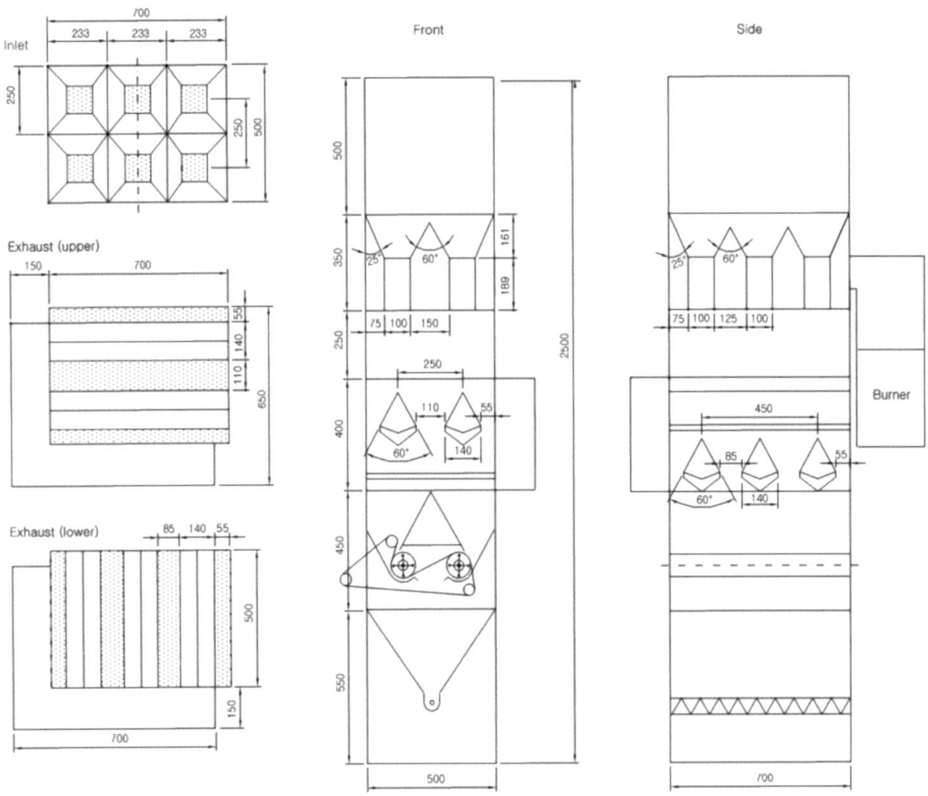

Figure 6.
Technical drawing of pilot-scale concurrent-flow dryer.

centrifugal fan through five exhaust air ducts. Rapeseed flow rate is controlled by two variable speed discharge augers. Rapeseed is out the dryer by discharge augers. Then, the grains are circulated by bucket elevator from the top of the dryer and flow down the vertical drying chamber.

The dimensions of the pilot-scale concurrent-flow dryer are shown in detail in **Figure 6**.

4.2 Experimental design

During the experiment the drying air temperature and grain temperature are continuously measured by temperature sensors (Thermocouple T-type, Omega, USA). The total of 16 temperature sensors was arranged at necessary positions inside and outside the dryer (**Figure** 7). Data from sensors were transferred to data logger system. Two computers were used to record the temperature data from the data logger (Datascan 7327, UK).

In the plenum section, drying air temperature input was measured by three sensors No. 1–3. In exhaust air ducts, two sensors at upper ducts (No. 10–11) and two sensors at lower ducts (No. 12–13) were arranged to measure exhaust air temperature. In the drying chamber, six sensors (No. 4–9) were arranged at six positions in the cross-section of the drying chamber to measure the temperature of grains. In two discharge augers, two sensors (No. 14–15) were arranged above the two discharge augers to measure discharge rapeseed temperature (rapeseed temperature after drying). To measure ambient air temperature, one temperature sensor (No. 16) was arranged outside the dryer (**Table 2**).

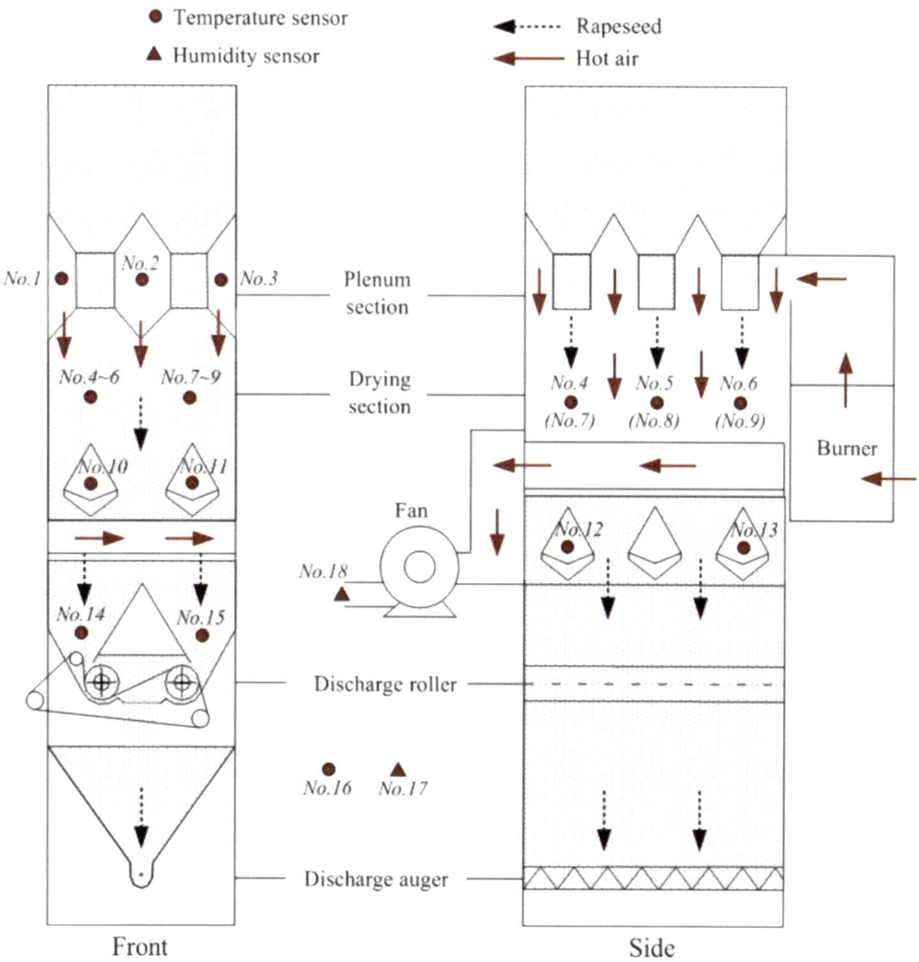

Figure 7.
Design layout for the temperature and relative humidity sensor locations in the pilot-scale concurrent-flow dryer.

Temperature measurement			**Sensors**
Drying air temperature	Inlet	3	No. 1–3
	Outlet	4	No. 10–13
Rapeseed temperature	Drying chamber	6	No. 4–9
	Discharge	2	No. 14–15
Ambient air temperature		1	No. 16
Relative humidity measurement			
Ambient air relative humidity		1	No. 17
Exhaust air relative humidity		1	No. 18

Table 2.
Temperature sensors distribution and hygrometer used in the experiment.

Two hygrometers (MTH4100, Sanyo, UK, 10 – 99%, ±2.5%) were installed to record the ambient air relative humidity (No. 17) and exhaust air relative humidity (No. 18) (**Table 2**).

In both experiments, grain flow velocity was set up at 5 m/h; this value is equivalent to the mass of circulated rapeseed which is 1000 kg/h. The grain flow velocity is controlled by two discharge rollers. The rotation of discharge rollers was 3.5 rpm. The rotation of discharge rollers was controlled by an inverter (S500, Mitsubishi, Japan).

A centrifugal suction fan 1 HP with air flow rate 30 cmm/m^2 [12] was used for sucking the exhaust drying air. An anemometer (Velocicalc-Plus, TSI, USA) was used to measure the drying air velocity.

The jet burner using kerosene (OL-3, Daewon, Korea) was used for heating drying air. The burner can be raised to the temperature of drying air up to 140°C. A temperature sensor (PT-100 Ω) was installed at the influx duct to control the burner, and a temperature control equipment (HSD-V2, Hansung, Korea) was used. An electric balance (A-200, Cass, Korea, accuracy 0.01 kg) was used to weigh the mass of Kerosene loss by drying process.

The dimension of drying chamber (height × length × width) is 0.5 m × 0.7 m × 0.5 m. Height of tempering section is 0.5 m. In both experiments, rapeseed samples using are Spring rapeseed, variety Sunmang F1-hybrid, were harvested in June in Jeonnam-do, Yeonggwang-gun. The samples of 200 kg were cleaned and stored in a refrigerator at a temperature of 4°C [13]. The initial moisture content of samples in Test 1 is 23.0% and in Test 2 is 23.2% (**Table 3**).

4.3 Experimental results

In both experiments, there are a difference in drying air temperature in the plenum chamber. The drying air temperature is highest at the position in front of the plenum, and lowest at back of plenum. In Test 1, the average temperature of drying air in the plenum section is 96.9, 84.2, and 83.9°C at the front, middle, and back of the plenum, respectively. In Test 2, the average temperature of drying air in plenum section is 128.1, 111.3, and 106.1°C at front, middle, and back of plenum, respectively (**Table 4**). The average temperature of drying air in the plenum chamber during drying process is 89.4 and 116.8°C for Test 1 and Test 2, respectively (**Figure 8**).

The temperature of rapeseed during drying at discharge augers (**Figure 9**) and the temperature of air at exhaust ducts (**Figure 10**) for both Test 1 and Test 2 are fairly uniform. Detailed drying conditions in Test 1 and Test 2 shown in **Table 5** and results of rapeseed drying in a pilot-scale dryer were summarized in **Table 6**.

Rapeseed	Test 1	Test 2
Initial weight (kg)	200	200
Initial moisture content (% w.b.)	23.0	23.2
Initial grain temperature (°C)	22.8	24.7

Table 3.
Initial rapeseed conditions.

Test no.	Drying air temperature (°C)		
	Front	Middle	Back
Test 1	96.9	84.2	83.9
Test 2	128.1	111.3	106.1

Table 4.
Drying air temperature in the plenum chamber.

After the experiment was completed, dried rapeseed samples were sealed in double-layer polythene bags for 24 h to reach ambient conditions [14]. The samples were then tested for germination. The germination tests were conducted according

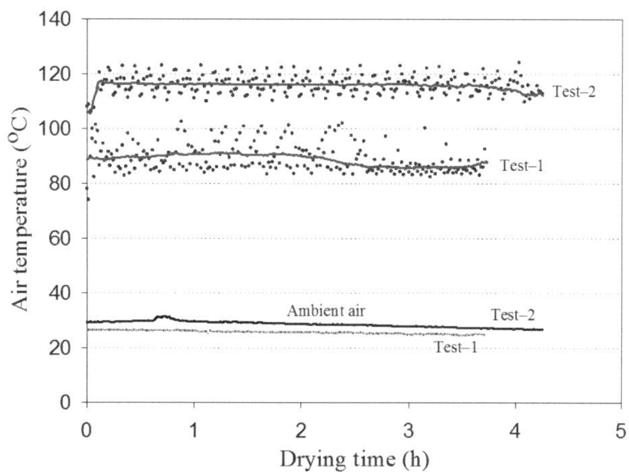

Figure 8.
Average variation of drying air temperature during drying process.

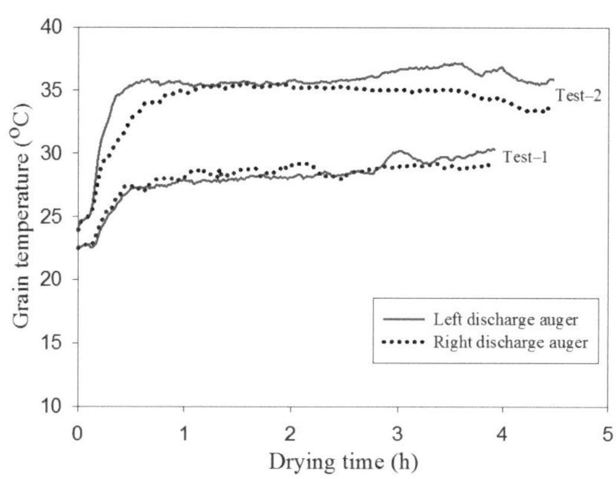

Figure 9.
Rapeseed temperature during drying process at discharge augers.

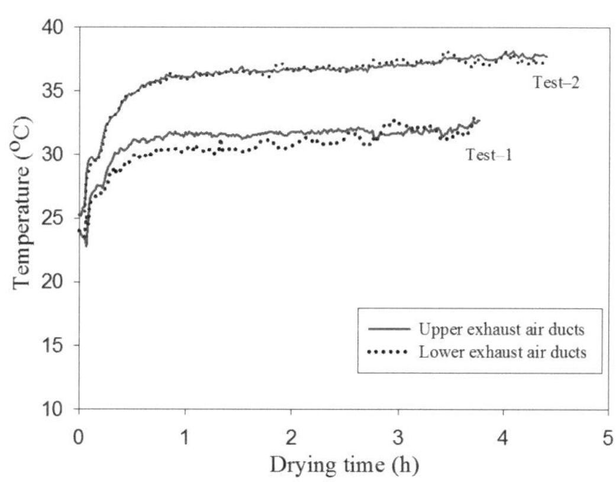

Figure 10.
Temperature of exhaust drying air in test 1 and test 2.

to protocols described for the standard germination test [15]. The average germination percentages of Test 1 and Test 2 are 94.7 and 84.5%, respectively [16].

4.4 Model validation

The simulation program was validated by comparison results of numerical model with experimental data of pilot-scale concurrent-flow dryer as described. The input data of simulation program were entered in accordance with the data of the actual experiment, such as initial rapeseed conditions, dryer specification, and drying air and ambient air conditions. The fitness of simulated results with measured results was evaluated based on the coefficient of determination (R^2) and the root mean square error (RMSE).

The R^2 of moisture content versus drying time in Test 1 and Test 2 were 0.994 and 0.997, respectively. The RMSE of moisture content in Test 1 and Test 2 were 0.334 and 0.506%w.b., respectively. The comparison of the measured and

Test No.	Average drying air temperature (°C)	Air flow rate (cmm/m^2)	Ambient air temperature (°C) Ave. (min–max)	Ambient relative humidity (%) Ave. (min–max)
Test 1	89.4	30	25.4 (24.2–26.4)	71.6 (67.1–75.3)
Test 2	116.8	25	28.6 (26.7–31.4)	63.4 (60.8–68.0)

Table 5.
Drying conditions for drying tests.

Experimental results	Test 1	Test 2
Initial moisture content (% w.b.)	23.0	23.2
Final moisture content (% w.b.)	13.8	11.4
Drying time (h)	3.90	4.22
Drying rate (% w.b./h)	2.38	2.80
Fuel consumption (kJ/kg-water)	4915	4831
Initial germination rate (%)	94.7	84.5

Table 6.
The results of rapeseed drying in pilot-scale dryer.

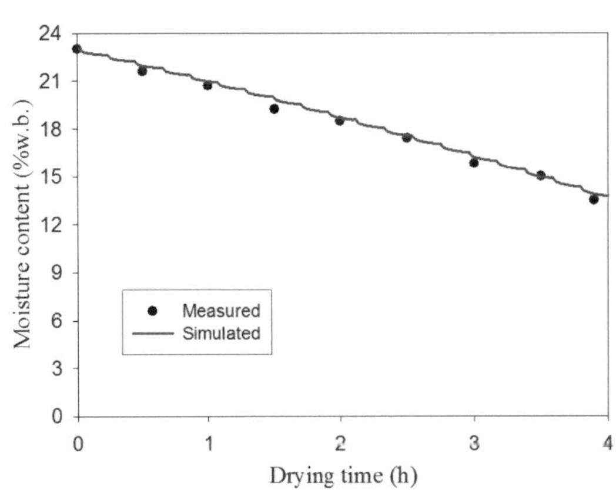

Figure 11.
Measured and simulated moisture content in test 1.

simulated moisture content during drying process for Test 1 is shown in **Figure 11** and for Test 2 shown in **Figure 12**.

The analytical results showed the good fitness between simulated moisture content and measured moisture content for both Test 1 and Test 2. This result showed the good agreement of the simulation program for predicting the moisture content of rapeseed in concurrent-flow dryer.

The simulated temperature of rapeseed during drying process has a good correlative with the experimental data. The R^2 of rapeseed temperature are 0.904 and 0.925 in Test 1 and Test 2, respectively. The RMSE of rapeseed temperature are 1.15 and 1.77°C in Test 1 and Test 2, respectively. The comparison of the measured and simulated temperature of rapeseed for Test 1 is shown in **Figure 13** and for Test 2 shown in **Figure 14**. The analytical results showed that simulated values have a very good fitness to measured values by experiment.

The discharge rapeseed temperature of the simulation program tends to be higher than the measured values during drying process. However, the average differences between measured and simulated values are small. It showed a good fitness between the values of the model and the values of the experiments.

The drying time, drying rate, fuel energy consumption, and germination ratio of rapeseed after drying were investigated. The comparison of the measured and simulated results of both Test 1 and Test 2 was listed in **Table 7**.

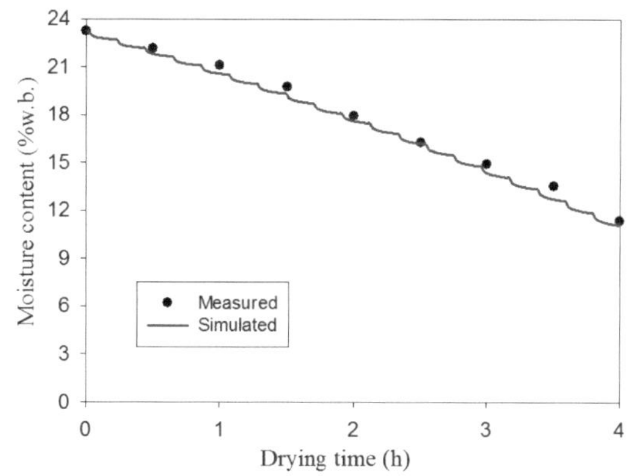

Figure 12.
Measured and simulated moisture content in test 2.

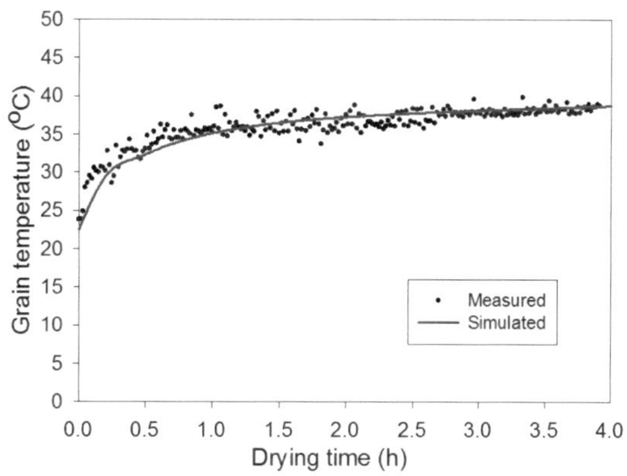

Figure 13.
Measured and simulated temperature of rapeseed in test 1.

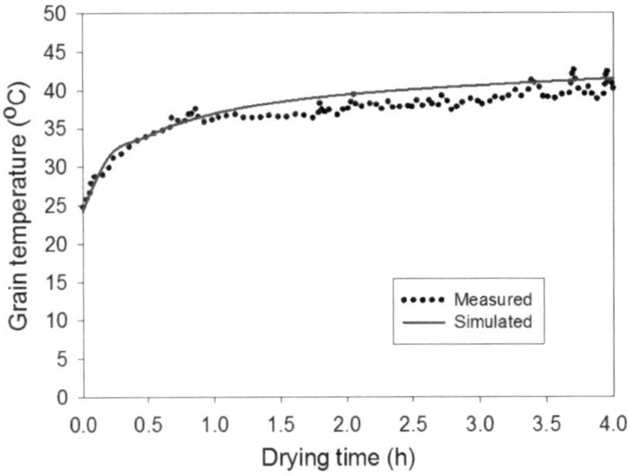

Figure 14.
Measured and simulated temperature of rapeseed in test 2.

Test 1	Measured	Simulated	Difference (%)
Final moisture content (% w.b.)	13.8	13.5	2.2
Drying rate (% w.b./h)	2.38	2.26	5.04
Fuel energy (kJ/kg-water)	4915	5153	4.62
Drying time (h)	3.9	4.2	7.14
Germination ratio (%)	94.7	96.5	1.87
Test 2	**Measured**	**Simulated**	**Difference (%)**
Final moisture content (% w.b.)	11.4	10.8	5.17
Drying rate (% w.b./h)	2.80	2.95	5.08
Fuel energy (kJ/kg-water)	4831	4417	8.57
Drying time (h)	4.22	4.2	0.47
Germination ratio (%)	84.5	92.5	8.64

Table 7.
Measured and simulated results of test 1 and test 2.

The difference between measured value and simulated value of final moisture content were 2.2 and 5.17%; drying rate were 5.04 and 5.08%; drying time were 7.14 and 0.47%; and germination ratio were 1.87 and 0.47%. The simulated values of fuel energy consumption for drying were 4.62 and 8.57% lower than the measured values for Test 1 and Test 2, respectively. The differences are derived from the sequential changes in drying air temperature, ambient temperature, and drying air humidity of the simulation model and experiment conditions. In general, there is a good fitness between measured values by experiment and simulated values from the simulation program for all of the output parameters such as final moisture content, drying time, drying rate, fuel energy, and germination ratio at the different experiment conditions.

5. Conclusions

Mathematical modeling for rapeseed drying on concurrent-flow dryer was built based on energy and mass transfer balances applied for both fluid and solid phases of a concurrent-flow dryer. Energy balances and mass balances are written on a differential volume located at an arbitrary location in the grain bed. The mathematical model consists of a set of four partial differential equations in four independent variables including air humidity, air temperature, grain temperature, and grain moisture content.

A computer simulation program for circulating concurrent-flow rapeseed dryer was developed using these models along with a detailed description of the program. The simulation program can predict the drying time, drying rate, drying air humidity and temperature, grain temperature and moisture content during drying process, water removal rate, drying fan parameters, germination ratio, fuel energy, and total energy consumption.

To evaluate a fitness of simulation program, a pilot-scale concurrent-flow dryer with a capacity of 200 kg/batch was designed, manufactured, and tested. Two drying experiments were conducted. The output parameters of the simulation program were compared and analyzed with experiment data.

The RMSE of simulated moisture contents ranged from 0.334 to 0.506%w.b. with the coefficient of determinations ranging from 0.994 to 0.997. The RMSE of simulated rapeseed temperatures ranged from 1.15 to 1.77°C with the coefficient of determinations ranging from 0.904 to 0.925. The experimental drying rates were 2.38 and 2.80% w.b./h. The difference between simulated value and measured value of drying rate were 5.04 and 5.08%; drying time were 7.14 and 0.47%; and germination ratio were 1.87 and 0.47%. The simulated fuel energy consumption for drying were 4.62 and 8.57% lower than the experimental values.

The analytic results showed that the simulation results have good fitness with experimental data. So, the mathematical modeling and the simulation program were proved their reliability and were shown to be a convenient tool for simulation of rapeseed drying in circulating concurrent-flow dryer.

Author details

Le Anh Duc[1]* and Keum Dong Hyuk[2]

1 Nong Lam University, Ho Chi Minh City, Vietnam

2 SungKyunKwan University, Suwon, South Korea

*Address all correspondence to: leanhduc@hcmuaf.edu.vn

References

[1] Thompson TL, Peart RM, Foster GH. Mathematical simulation of corn drying—a new model. Transactions of ASAE. 1968;**24**(3):582-586

[2] Felipe CAS, Barrozo MAS. Drying of soybean seeds in a concurrent moving bed: Heat and mass transfer and quality analysis. Drying Technology. 2003; **21**(3):439-456

[3] Keum DH, Han JG, Kang SR, Kim OW, Kim H, Han JW, et al. Development of rice circulating concurrent-flow dryer (I) - performance teset of pilot scale dryer. Journal of Biosystems Engineering. 2005;**10**(2):97-106

[4] Han JW, Keum DH, Kim W, Duc LA, Cho SH, Kim H. Circulating concurrent-flow drying simulation of rapeseed. Journal of Biosystems Engineering. 2010;**35**(6):401-407

[5] Keum DH. Simulation of Agricultural Products and Foods Process Engineering. South Korea: SungKyunkwan University Publisher; 2005

[6] Duc LA, Han JW, Hong SJ, Choi HS, Kim YH, Keum DH. Physical properties of rapeseed (I). Journal of Biosystems Engineering. 2008;**33**(2):101-105

[7] Gallaher GL. A method of determining the latent heat of agricultural crops. Agricultural Engineering. 1951;**32**(1):34-38

[8] Duc LA, Hyuk KD. Equilibrium moisture content isotherm characteristics of rapeseed. Asia Pacific Journal of Sustainable Agriculture, Food and Energy. 2016;**4**(1):10-14

[9] Cenkowski S, Muir WE, Jayas DS. Simulation of canola and barley drying in deep bed. Journal of Food Process Engineering. 1989;**12**:171-190

[10] Duc LA, Han JW, Keum DH. Thin layer drying characteristics of rapeseed (*Brassica napus* L.). Journal of Stored Products Research. 2011;**47**(1):32-38

[11] Fausett LV. Applied Numerical Analysis Using Matlab. Upper Saddle River, New Jersey: Prentice Hall Inc.; 1999. p. 07458

[12] Hong SJ, Duc LA, Han JW, Kim H, Kim YH, Keum DH. Physical properties of rapeseed (II). Journal of Biosystems Engineering. 2008;**33**(3):173-178

[13] Cassells JA, Caddick LP, Green JR, Reuss R. Isotherms for Australian canola varieties. In: Proceedings of the Australian Postharvest Technical Conference. 2003. pp. 59-63

[14] ANSI/ASAE S448.1. Thin-layer drying of agricultural crops. In: ASAE Standards 51st Edition. 2004. pp. 598-600

[15] Association of Official Seed Analysis. Rules for testing seeds. Journal of Seed Technology. 1993;**16**:1-113

[16] Duc LA, Han JW. The effects of drying conditions on the germination properties of rapeseed. Journal of Biosystems Engineering. 2009;**34**(1): 30-36

Chapter 5

Determination on Fluidization Velocity Types of the Continuous Refined Salt Fluidized Bed Drying

Bui Trung Thanh and Le Anh Duc

Abstract

After the centrifugation stage, refined salt particles have rather high moisture content; therefore, the moist salt particles in contact with each other will stick together in a short time. In particular, the moist salt particles will stick together faster and tighter and form a larger unit when they are exposed to drying hot air. For this reason, the refined salt was dried by rotary drum dryers with vibrating balls distributed along the drum or a vibrating fluidized bed dryers. These drying methods make poor product sensory quality, low product recovery efficiency, while also lead to an increase of heat and electricity energy consumption. In order to increase the efficiency of refined salt drying technology by conventional continuous fluidized bed dryers, the chapter focuses on the study of aerodynamic properties of refined salt grains in the continuous fluidized particle layer. The content of the chapter presents theoretical and empirical methods to determine fluidization velocity types in designing a continuous fluidized bed dryer.

Keywords: refined salt, solid particles, aerodynamic, minimum fluidization velocity, homogeneous fluidization, bed fraction, fluidized bed dryer

1. Introduction

1.1 Preface

The phenomenon in which solid particles float in a gas stream and have a liquid-like property is called a fluidized bed. This phenomenon of fluidization in gas or liquid flow was discovered by Fritz Winkler in the 1920s [1]. This one was investigated by Lewis et al. and had been raised to fluidized theory [2]. The first commercial fluidized bed dryer was installed in USA in 1948 [3]. The fluidized bed technology is used for drying of bulk materials, includes examples such as: vibrating fluidized bed dryer, normal fluidized bed dryer without vibrating device and pulsed fluidized bed dryer.

Mathematical modeling and computer simulation of grain drying are now widely used and become an important tool for designing new dryers, for analyzing existing drying systems and for identifying drying conditions [4]. Identifying the drying conditions is necessary to establish the optimal protocol for ensuring seed quality [5]. To solve the simulation models, equations concerning aerodynamic properties such as the gas stream velocity and particle velocity are the most important components. The aerodynamic properties are affected by shape, density and size of particles [6].

Order	Function of gases	Function of solid materials	Equipment in practice
1	Heat carrier	Materials do not react to gases	• Heaters supply to materials • Heat exchangers with heat recovery
2	Loading and transportation of particle materials	Materials do not react to gases	Pneumatic solid particle carrier
3	Heat carrier and material loading and transportation	Materials do not react to gases	Aerodynamic dryers
4	Chemical agents	• Chemical reaction • Activated material for chemical reactions • As inert material which is used in fluidized bed combustion of boiler	• The burner burns • Gasification equipment • Catalytic reconstitution • Oxidation equipment • Processing of metallurgical surfaces and annealing furnace
5	Agitation	Materials do not react to gases	Mixer of different materials in the tank

Table 1.
Relations and functions of gas interacting with solid particle in real production.

In recent years, the fluidized bed technology has been concerned in the application in the sugar drying and refined salt drying in Vietnam. In the scope of this chapter, we discuss the issues related to the hydrodynamics of refined salt particles in the gas stream at ambient temperature and temperature equivalent to that of drying particle. The focus of this chapter is to determine the velocity values of the gas through the particle layer to form the minimum, homogeneous and critical fluidized layers.

In industrial manufacturing practice, we often encounter the contact, interaction between solid-particle materials and gases. These interaction phenomena are described in **Table 1**.

1.2 Applied materials in fluidized bed drying technology

According to Abrahamsen and Geldart [7], the two most important factors affecting the fluidization characteristic of the particle layer are particle size and particle density.

Geldart [8] visually observed the various conditions and classified fluidizable particles into four groups: A, B, C and D. As a result, the classification was related to the influence of the average particle size and particle density on the properties of the fluidized layer, as depicted in **Figure 1** and **Table 2**.

According to the content of this chapter, we focus on approaching, researching and experimenting on the mechanism and principle of interaction between air and the applied material in fluidized bed drying. The approach method is to arrange a stream of heat-carrying air blowing from the bottom of the particle chamber through a gas distributor (the holes arranged at an angle to the cross section of the tank). Hot air stream is evenly distributed and touches the surface of particles in tank (the particle layer was on the gas distributor). The continuous air stream ensures that the contact of particle surfaces with the gas flow is consecutive. The nature of the gas flowing through the particle layer may be laminar, turbulent, or transition flow at the material contact surface. The inflow of hot air affects the velocity of the interaction between the gas stream and the material.

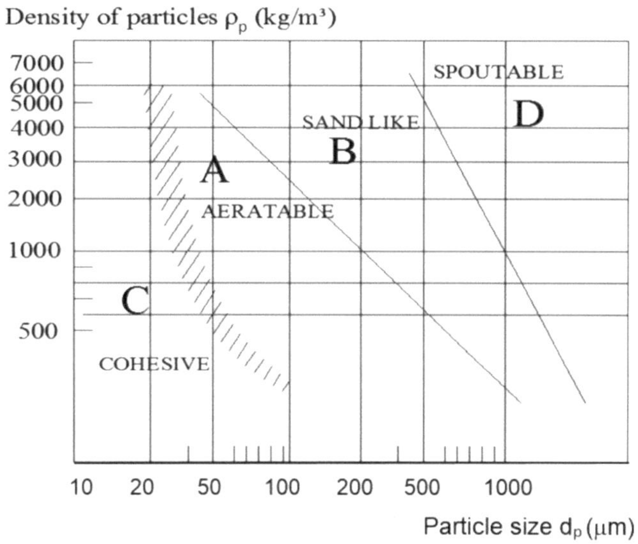

Figure 1.
Diagram of the Geldart classification of particles [9, 10].

1.3 Basic concept of fluidized particle layer and principle of fluidized particle layer creation

When the material layer is fluidized, its state is converted from a fixed bed to a dynamic state. The particle layer has liquid-like properties. The surface area of the particles contacting with the fluid increases, and therefore the heat transfer ability from fluid to particles rapidly rises.

In order for the fluidization phenomena to occur in the bulk material layer, the air stream must have sufficient pressure and velocity. The air stream flows upward, passes through the particle materials (follow the linear increment) through uncountable air holes of the distributor, which are arranged at the bottom of the particle material layer. When the velocity of air stream is small, pressure exerted on particles is small, the particle layer maintains its original fixed bed (the state from 0 before point A, **Figure 2**). As the air velocity is increased further, the aerodynamic traction appears, which has opposite effect of the gravitational force of the particles, causing the expansion of particle layers in a volume, and the particles begin to move apart from each other (at point A, **Figure 1**).

By further raising the velocity of air stream to the critical value, the friction force between the particles and air is equal to the weight of particles. At this time, the vertical component of the compression pressure is eliminated, the upward-pulling force equals the downward gravity, causing the particle material to be suspended in the air stream. When the gas velocity reaches the critical value, the particle material layer will be converted to complete fluidization state, called the fluidization particle layer and having the liquid-like properties (position from A to B in **Figure 2**).

By further increasing the velocity of the gas, the bulk density of the particle layers continues to decrease, its fluidization becomes more violent, until the particles no longer form a bed and are swept up and fallen down in the fluidization motion (position from B to C, **Figure 2**). At this time, each particle material is covered with gas flow, the intensity of the heat and material transfer occurs violently. When granular material is fully fluidized, the bed will conform to the volume of the chamber, its surface remaining perpendicular to gravity; objects with a lower

Particle group characteristics	C group	A group	B group	D group
Particle size (μm)	0–3	$30 \le d_p \le 100$	$100 \le d_p \le 1000$	≥ 1000
Density (kg/m^3)	Smallest	1400	400–4500	Lower than other materials
The most obvious characteristics of group	• Particles are cohesive and linked • Difficult fluidization	• Easy fluidization • Dense phase expands stably before bubbling starts	• Starting on bubble creation at the minimum fluidized velocity value	• The rough solid particles
Typical granulars	• Flour • Cement	• Milk flour • FCC granular	• Construction sands	• Pebbles in rice • Coffee beans, wheat, lead shot
Properties of fluidization layer				
Particle layer expansion	• Channeling possibilities in the particles layer easily	• The large bed expansion before bubbling is started • The minimum fluidization velocity is smaller than minimum bubbling	Medium	Difficult to fluidize evenly (low)
Properties of air bubbles	• Do not form bubble	• It does not form bubble fluidization • There is a maximum bubble size	• The bubbles rise faster than the interstitial gas • Bubbles are large and grow rapidly and coalescence, as they rise through the bed	• Bubbles rise more slowly than the rest of the gas percolating through the emulsion
Property of mixed solid particles	Very low	High	Solids' recirculation rates are smaller	Low
Spaying	None	None	Only occurs in the upper layer	• It occurs • Only occur in under layers

Table 2.
The classification of fluidization properties of particle groups according to Geldart [9, 11].

density than the bed density will float on its surface, bobbing up and down, while objects with a higher density sink to the bottom of the bed.

Fluidization has many applications in many technologies of manufacturing practice, such as mixing different types of granular materials; fluidized bed drying; cooling grain after drying; supporting interaction between chemicals in the fluidized bed; granulation technology; film coating technology of medicine and pharmacy; manufacturing technology through the combined use of organic and inorganic fertilizers; and biomass fuel combustion technology in fluidized bed.

To clarify the dynamics of the fluidized beds for application in refined salt drying in the fluidized bed, the theoretical and experimental issues determining the velocity of gas through the particle layers to form different fluidized layers are described as follows.

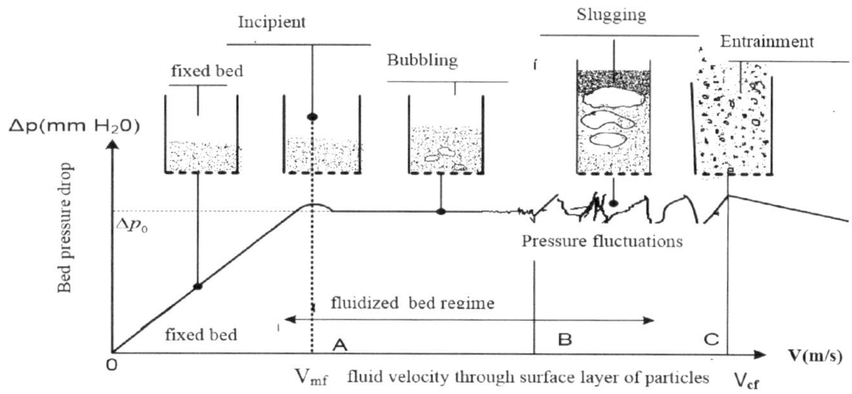

Figure 2.
Particle layer states with gas velocity changing [10].

2. Methodology

2.1 Determination of the minimum fluidization velocity from publications

2.1.1 Determination of V_{mf} using the Ergun equation

When the air stream having sufficient pressure and velocity passes through the static spherical particle layers, they begin to expand (the particles become "flexible"). In this condition, it is called the minimum particulate fluidization state and is described by the modified equation of Ergun (see position A in **Figure 1**) [12].

$$\frac{\Delta P}{H_{mf}} = \frac{150\mu_f(1-\varepsilon_{mf})^2 V_{mf}}{\varepsilon_{mf}{}^3 d_p^2} + \frac{1.75(1-\varepsilon_{mf})\rho_f V_{mf}{}^2}{\varepsilon_{mf}{}^3 d_p} \tag{1}$$

For particles of arbitrary shapes, the pressure drop of air stream at the minimum fluidization state is represented by Eq. (2). The spherical value of particle material got in the Eq. (1) [10] is given by:

$$\frac{\Delta p}{H_{mf}} = 150\frac{(1-\varepsilon_{mf})^2}{\varepsilon_{mf}{}^3}\frac{\mu_f V_{mf}}{(\phi d_m)^2} + 1.75\frac{(1-\varepsilon_{mf})}{\varepsilon_{mf}{}^3}\frac{\rho_f V_{mf}{}^2}{\phi d_m} \tag{2}$$

For the particle layer to be converted from a fixed state to a fluidization state, the pressure of air stream must be large enough to overcome the weight of the particle layers and it is determined by Eq. (3) [10].

$$\Delta p = \frac{m}{\rho_p A}\left(\rho_p - \rho_f\right)g \tag{3}$$

In Eq. (3), it is considered that there was no interaction force between particles in layer and no interaction between particles and the wall of the tank. So, that did not cause the pressure increasing effect. Thus, the pressure drop of the air stream was constant while increasing the gas velocity from the smallest fluidization velocity to the value when the entrainment process of particles occurred (position C, **Figure 2**).

In Eqs. (1) and (2), it is also shown that the pressure drop of the gas stream that is generated through the fluidized particle layer is depended on the particle size (d_p), the bed voidage (ε) and the gas temperature ($t^\circ C$). According to Eq. (3), the pressure drop of the gas stream through the particle layers is dependent upon the material mass (m), the gas distributor grate area (A), particle density (ρ_p) and gas density (ρ_f). Thus, we could calculate the minimum fluidization velocity value (V_{mf}), which was based on Eqs. (1)–(3) for non-spherical particles by solving Eq. (4).

$$\frac{m.H_{mf}}{\rho_p.A}\left(\rho_p-\rho_f\right)g=150\frac{\left(1-\varepsilon_{mf}\right)}{\varepsilon_{mf}{}^{3}}\frac{\mu_f V_{mf}}{\left(\phi d_p\right)^2}+1.75\frac{1}{\varepsilon_{mf}^{3}}\frac{\rho_f V_{mf}{}^{2}}{\phi d_p} \tag{4}$$

The minimum fluidization velocity (V_{mf}) is the root of Eq. (4), which is based on available parameters, such as the height of minimum fluidization bed (H_{mf}), the mass of particles in the air distributor (m), the area for gas distribution or called cross-sectional area of the bed (A), particle density (ρ_p), air density (ρ_f), mean particle diameter (d_p), spherical degree of particles (ϕ) and void fraction at minimum fluidization particle layer (ε_{mf}).

Commonly, the sphere degree of particles (ϕ) must be determined in experiments [10]. The spherical degree of refined salt particles was found out by Bui [13–15]. Theoretically, in order to fluidize the particle layer, the actual weight of the solid particles must be equal to the force exerted on the particle layers and that is equal to the pressure drop across the bed (ΔP) multiplied by the cross-sectional area of the chamber (A). A minimum fluidization layer which must have determined layer thickness (H_{mf}), void fraction (ε_{mf}), then the expanded volume of the fluidized particles (U) has the value of the Eq. (5):

$$U=(1-\varepsilon_{mf}).A.H_{mf} \tag{5}$$

And the actual gravity of the particle mass has a value of:

$$W=(1-\varepsilon_{mf})\left(\rho_p-\rho_f\right).A.H_{mf.}.g \tag{6}$$

The balance of the real gravity components of the particle mass and the upward force exerted on the particle mass of the gas flow was calculated according to

$$\Delta P=(1-\varepsilon_{mf})\left(\rho_p-\rho_f\right).H.g \tag{7}$$

Substituting Eq. (8) into Eq. (1) or Eq. (2) yields Eq. (9).

$$\left(1-\varepsilon_{mf}\right)\left(\rho_p-\rho_f\right).H_{mf}.g=150\frac{\left(1-\varepsilon_{mf}\right)^2}{\varepsilon_{mf}{}^{3}}\frac{\mu_f V_{mf}}{\left(\phi d_p\right)^2}+1.75\frac{\left(1-\varepsilon_{mf}\right)}{\varepsilon_{mf}{}^{3}}\frac{\rho_f V_{mf}{}^{2}}{\phi d_p} \tag{8}$$

$$150\frac{\left(1-\varepsilon_{mf}\right)}{\phi^2\varepsilon_{mf}^{3}}\mathrm{Re}_{mf}+\frac{1.75}{\phi\,\varepsilon_{mf}^{3}}\mathrm{Re}_{mf}^{2}=Ar \tag{9}$$

Giving physical parameters of particle and gas into the Eq. (8), velocity (V_{mf}) was found out. In case of very small particles, the gas stream regime through the particle layer was laminar flow and the minimum fluidization velocity should be calculated by the Ergun equation [12]. In case of $Re_{mf} < 1$, we use Eq. (9) to calculate the minimum fluidization velocity of gas.

2.1.2 Determination of V_{mf} by the correlation of Re_{mf} and Archimeter (Ar)

When gas passes through the particle bulk, which can have any shape, the minimum fluidization Reynolds coefficient (Re_{mf}) is determined by Eq. (10).

$$\mathrm{Re}_{mf} = \frac{\rho_{mf}.V_{mf}.d_p.\phi}{\mu_f} \tag{10}$$

Eq. (11) describes the correlation between Ar and Re_{mf} with void fraction at minimum fluidization particle layer (V_{mf})

$$Ar = 150\frac{(1-\varepsilon_{mf})}{\phi^2\,\varepsilon^3_{mf}}\mathrm{Re}_{mf} + \frac{1.75}{\phi\varepsilon^3_{mf}}\mathrm{Re}^2_{mf} \tag{11}$$

Archimeter (Ar) is determined by Eq. (12) for particles of any shape.

$$Ar = \frac{\rho_f\left(\rho_p-\rho_f\right)g\,(\phi d_p)^3}{\mu_f^2} \tag{12}$$

Set up Eq. (13):

$$\frac{\rho_f\left(\rho_p-\rho_f\right)g\,(\phi d_p)^3}{\mu_f^2} = 150\frac{(1-\varepsilon_{mf})}{\phi^2\,\varepsilon^3_{mf}}\mathrm{Re}_{mf} + \frac{1.75}{\phi\,\varepsilon^3_{mf}}\mathrm{Re}^2_{mf} \tag{13}$$

Substituting physical parameters into the Eq. (12), which includes the particle density (ρ_p), the gas density at the temperature of minimum fluidization velocity (ρ_f), the air dynamic viscosity (μ_f), spherical degree of particle (ϕ), void fraction at the minimum fluidization velocity (ε_{mf}), mean particle diameter (d_m) [13–15] and getting Ar number value into the Eq. (13). Then, we solved the quadratic equation to find out the root of equation Re_{mf} in Eq. (10), we got only the positive value. Thus, we calculated the minimum fluidization velocity (V_{mf}) from Eq. (10).

2.1.3 Determination of V_{mf} by the Kozeny-Carman correlation

Kozeny-Carman gave the formula of calculation of the minimum fluidization velocity for a very small particle size with the $\mathrm{Re}_{mf} < 10$ in Eq. (14) described in Yates [16].

$$V_{mf} = \frac{g\left(\rho_p-\rho_f\right)}{150\mu_f}\frac{\varepsilon_{mf}{}^3}{1-\varepsilon_{mf}}\phi^2 d_p^2 \tag{14}$$

For spherical particles or sphericity equivalent, the bed voidage of the minimum fluidization $\varepsilon_{mf} = 0.4 \div 0.45$.

2.1.4 Determination of V_{mf} by correlation of Wen and Yu

In case of the unavailability of the sphericity of particles, we determine the minimum fluidization velocity (V_{mf}) by using of the experimental correlation of Wen and Yu [17]. An empirical formula of calculation of the void fraction at minimum fluidization particle layers in Eq. (15) or Eq. (16) with the available sphericity degree of particle (ϕ) or the calculation of the sphericity degree of particle (ϕ) in case void fraction (ε_{mf}) at minimum fluidization particle layer is available, which was also described by Wen and Yu equation (cited in Howard, 1989) [10].

$$\frac{1-\varepsilon_{mf}}{\phi^2\,\varepsilon_{mf}^3} \approx 11 \text{ or } \frac{1}{\phi\,\varepsilon_{mf}^3} \approx 14 \tag{15}$$

We can use the calculation of the void fraction at minimum fluidization particle layers from the other correlation, which is also converted from Wen and Yu.

$$\varepsilon_{mf} = \left(\frac{0.071}{\phi}\right)^{1/3} \tag{16}$$

It is based on the calculation of the void fraction (ε_{mf}) (at position A in **Figure 2**) of Eq. (15) or Eq. (16) and substituting the obtained ε_{mf} value into the Ergun Eq. (2), we have Eq. (17).

$$Ar = 1650\ \mathrm{Re}_{pmf} + 24.5\ \mathrm{Re}_{pmf}^2 \tag{17}$$

Using the calculated Ar number from Eq. (12) and substituting the Ar value into Eq. (17), we obtain Eq. (18) to calculate the particle Reynolds number at the minimum fluidization velocity (Re_{mf}).

$$\mathrm{Re}_{mf} = \frac{-1650 \pm \{1650 + (4 \times 24.5\ Ar)\}^{1/2}}{2 \times 24.5} \tag{18}$$

Taking the positive square root, we get Eq. (19):

$$\mathrm{Re}_{mf} = \left(33.7^2 + 0.0408\ Ar\right)^{1/2} - 33.7 \tag{19}$$

It was applied to calculate for solid particles with size larger than 100 μm [10]. From the Re_{mf} value that was found out in the Eq. (19), V_{mf} is calculated according to Eq. (10). In case of solid particles with small size (C group of Geldart, 1973) in the specified temperature conditions, the V_{mf} value is calculated in Eq. (20) by Wen and Yu.

$$V_{mf} = 7.90.10^{-3} d_p^{1.82}\left(\rho_p - \rho_f\right)0.94\mu_f^{-0.83} \tag{20}$$

2.1.5 Determination of V_{mf} by the correlation of Beayens and Geldart

For solid spherical particles with diameters ranging from 0.05 to 4 mm (0.05 mm < d_p < 4 mm) and particle density ranging from 850 to 8810 kg/m^3 (850 kg/m^3 < ρ_p < 8810 kg/m^3), the method of calculation of V_{mf} was proposed by Beayens and Geldart as shown in Eq. (21) [18].

$$Ar = 1823\ \mathrm{Re}_{mf}^{1.07} + 21.7\ \mathrm{Re}_{mf}^2 \tag{21}$$

Then the V_{mf} can be calculated from Eq. (22) in case of available solid particle and gas parameters.

$$V_{mf} = \frac{9.125 \times 10^{-4}\left(\left(\rho_p - \rho_f\right)g\right)^{0.934} d_p^{1.8}}{\mu_f^{0.87}\rho_p^{0.66}} \tag{22}$$

2.1.6 Determination of V_{mf} by correlation of Goroshko

The minimum fluidization velocity of spherical particles was determined by correlation shown in Eq. (23) by Goroshko described in Howard [10, 19].

$$a\ \mathrm{Re}_{mf}^{2} + b\ \mathrm{Re}_{mf} - Ar = 0 \tag{23}$$

where

$$a = \frac{1.75}{\phi\,\varepsilon_{mf}^{3}}\ \text{and}\ b = \frac{150\,(1-\varepsilon_{mf})}{\phi^{2}\,\varepsilon_{mf}^{3}} \tag{24}$$

We get ϕ is 1.0 (ϕ = 1) then solving the Eq. (23) take the spherical degree value, we have the Eq. (25).

$$\mathrm{Re}_{mf} = \frac{-b + \left(b^{2} + 4aAr\right)^{1/2}}{2a} \tag{25}$$

Multiplying $\left(b + \sqrt{b^{2} + 4aAr}\right)$ by the numerator and denominator of Eq. (25), we have the Eq. (26)

$$\mathrm{Re}_{mf} = Ar\left\{\frac{b}{2} + \left[\left(\frac{b}{2}\right)^{2} + aAr\right]^{1/2}\right\}^{-1} \tag{26}$$

And its value was determined by Eq. (27) by Goroshko:

$$\mathrm{Re}_{pmf} = \frac{Ar}{b + \sqrt{aAr}} \tag{27}$$

Eq. (27) is different from Eq. (26) by the added value $\left(\frac{b}{2}\right)^{2} + aAr = \frac{b}{2} + \sqrt{a.Ar}$. There is a difference in Re_{mf} value between the Goroshko equation and Ergun equation. The Re_{pmf} of Goroshko equation [Eq. (27)] is smaller than Re_{mf} of Ergun [Eq. (26)]. This deviation interval depends on the value of Archimeter (Ar). Thus, we have a correlation equation, which is described in Eq. (28).

$$\frac{\mathrm{Re}_{mf}\,(Ergun)}{\mathrm{Re}_{pmf}\,(Goroshko)} = \frac{b + \sqrt{aAr}}{\left\{b/2 + \left[(b/2)^{2} + aAr\right]^{1/2}\right\}} \tag{28}$$

2.1.7 Determination of V_{mf} following Goroshko and Todes equation

The minimum fluidization velocity (V_{mf}) of sphericity particles was defined from the Re_{mf} by Goroshko et al. in Eq. (29) [19].

$$\mathrm{Re}_{pmf} = \frac{Ar}{150\,\frac{1-\varepsilon_{mf}}{\varepsilon_{mf}{}^{3}} + \sqrt{\frac{1.75}{\varepsilon_{mf}^{3}}Ar}} \tag{29}$$

In case of non-spherical particles with different sizes, the Re_{pmf} value was error from 15–20% in case we use the Eq. (29) of calculation described in [20]. In the case of rapid calculation, we considered the bed voidage of the minimum fluidization state to be equal to the bed voidage at static particle layers ($\varepsilon_o = \varepsilon_{mf} = 0.4$), and the Re_{mf} is calculated in Eq. (30) [21].

$$\mathrm{Re}_{mf} = \frac{Ar}{1400 + 5.22\sqrt{Ar}} \tag{30}$$

2.1.8 Determination of V_{mf} following Leva

From the formula of Carman-Kozan $k = \frac{g.\rho_f.\varepsilon_{mf}}{\mu_f k_c S}$ described in (as cited in Leva, [22]) yield the formula to define the minimum fluidization velocity [22, 23].

$$V_{mf} = \frac{5 \times 10^{-3} (\phi d_p)^2 \left(\rho_p - \rho_p\right) g {\varepsilon_{mf}}^3}{\mu_{mf} \left(1 - \varepsilon_{mf}\right)} \quad (31)$$

The Leva formula is used in case of Reynolds to be smaller than 10 ($Re_{mf} < 10$). In case of Reynolds to be larger than 10 ($Re_{mf} > 10$), there is an adjustment factor added into this formula.

2.1.9 Determination of V_{mf} according to Kunii-Levenspiel

The formula of Kunii-Levenspiel was simplified from the Ergun formula and it gave out two cases of calculation of the minimum fluidization velocity. In the first case for solid particles of small size with $Re_{mf} < 20$, we have to use Eq. (32).

$$V_{mf} = \frac{(\phi d_p)^2}{150 \mu_f} \frac{\left(\rho_p - \rho_f\right) g {\varepsilon_{mf}}^3}{\left(1 - \varepsilon_{mf}\right)} \quad (32)$$

We have to use Eq. (33) for the larger particle size with Reynolds number larger than 1000 ($Re_{mf} > 1000$).

$${V_{mf}}^2 = \frac{(\phi d_p) \left(\rho_p - \rho_f\right) g.{\varepsilon_{mf}}^3}{1.75 \rho_f} \quad (33)$$

2.1.10 Determination of V_{mf} based on the bed voidage problem

There is a correlation equation of particle mass balance at the minimum fluidization state (fluidization without bubbles), which was created by Kunii and Levenspiel as shown in Eq. (34).

$$g.H_0(1 - \varepsilon_0)\rho_p.A = \left(1 - \varepsilon_{mf}\right)\rho_p g.H_{mf} \quad (34)$$

Thus, we can obtain a correlation as shown in Eq. (35).

$$\frac{H_{mf}}{H_0} = \frac{(1 - \varepsilon_0)}{1 - \varepsilon_{mf}} \quad (35)$$

According to Ginzburg, described in [18], the bed voidage of minimum fluidization and height of particle layer are calculated by Eq. (36) and Eq. (37).

$$\varepsilon_{mf} = \varepsilon_0 \times 10\% \quad (36)$$

$$H_{mf} = H_0 \times 10\% \quad (37)$$

McCabe et al. proposed $\varepsilon_{mf} = 0.4 \div 0.45$ for the spherical particle [24]. The bed voidage of minimum fluidization particle layers was 0.5 ($\varepsilon_{mf} = 0.5$) for larger particle size. The bed voidage is equal to 1.0 ($\varepsilon_t = 1.0$) when the particle layers are attracted to the gas stream (see position C in **Figure 2**).

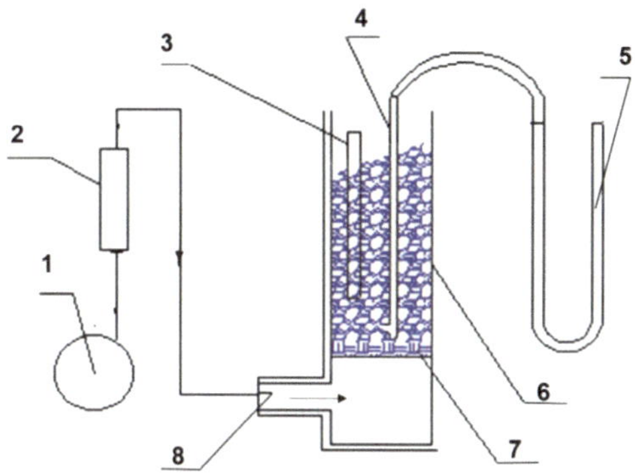

Figure 3.
Model for determination of the minimum fluidization velocity. 1. Centrifugal fan; 2. air heater; 3. thermometer for surface particle temperature measurement; 4. pitot tube for measurement of dynamic pressure and total pressure of air; 5. U-manometer; 6. chamber of fluidization; 7. air distributor; 8. drying air inlet.

2.2 Physical model of experiment

Experimental arrangement of determining the minimum fluidization velocity is shown in **Figure 3**.

In order to gradually increase the bed surface velocity of hot air via the particle layers, the air fan (1) is equipped with an inverter to change the rotation of the fan motor.

2.3 Experimental equipment

The instruments in experiments include a moisture analyzer (Axis AGS100, Germany), measurement error ± 0.01%; a digital electronic scale (Satorius MA45, Germany), measurement error ± 0.001 g; an air velocity meter (Extech SDL350 Taiwan), measurement error ± 0.01 m/s and a digital thermometer (WIKA CTH6300, Germany), measurement accuracy 0.001°C. This instrument has two measuring rate modes including fast at 4/s and slow at 1/s; an inclined manometer (T10, UK), measured range is 0–280 mmwg with error ± 0.1% and a pitot pipe (PT6300, 304 Germany), measurement range is 0–400 mmwg with error ± 0.1%. For measurement on the bulk density and density of refined salt particles, we used instruments such as Graduated pipet, buret, graduated cylinder, all of them made in Germany with error measurement ± 0.01 ml. The HCl acid is used for density measurement of refined salt particles.

No.	Equipment/parts	Technical parameter
1	Drying air fan	Flow: 0.63 m^3/s; total pressure: 1244 Pa; motor power: 2.2 kW
2	Electrical heater	Overall dimension (L × W × H): 600 × 630 × 275 mm; heating power: 1.0 kW; number of heater bars: 6
3	Drying chamber	Overall dimension (L × W × H): 1750 × 300 × 350 mm Fabrication material: SUS304
4	Salt dust settling chamber	Overall dimension (L × W × H): 1750 × 450 × 350 mm Fabrication material: SUS304

Table 3.
The basic parameters of the continuous fluidized bed dryer for experiment by authors.

2.4 The materials of refined salt particles

The material of refined salt particles was supplied by a combined hydraulic separating-washing-grinding machine in the saturated saltwater condition and which was dried by a continuous centrifugal machines. Samples of refined salt were randomly taken at different sizes at Vinh Hao salt company in Binh Thuan Province, Bac Lieu salt company and Sea salt Research Center of Vietnam for analysis (**Table 3**) [14].

3. Results and discussions

The above section presented nine methods to calculate minimum fluidization velocity based on the physical parameters of particles and physical thermal parameters of gas stream. These parameters were obtained from experiments in combination with the correlation calculation or empirical formulas.

In order to have the basis of comparison and accuracy evaluation of each calculating method in comparison with the empirical method, the theoretical calculation was carried out for refined salt particles with diameters of 1.5 mm, 1.2 mm, 0.9 mm, 0.6 mm and 0.3 mm. On the other hand, to achieve empirical result, samples of dried refined salt particles (of which mean-diameter was determined) were taken randomly from a combined hydraulic-separating-washing-crushing machine, presenting various particle sizes of the raw material that was put in the dryer.

3.1 Results of theoretical and empirical calculations for determining V_{mf} of refined salt particles

3.1.1 Some primary conditions for determining the V_{mf} by theoretical calculations

When calculating the pressure drop across a refined salt particle layer, we relied on the empirical results of physical parameters of particles and air (summarized in **Table 4**). Specific notes for each calculating method are as follows:

a. Calculation based on Ergun equations and correlations of pressure

Applying to calculate the minimum fluidization velocity (V_{mf}) for refined salt particles with the fixed bed height (H_0) is 30 mm, the bed voidage (ε_0) is 0.5 using Eqs. (35) and (36) to find out the minimum fluidization state including $H_{mf} = 1.1 \times H_0 = 33$ mm; bed voidage $\varepsilon_{mf} = 1.1 \times \varepsilon_0 = 0.56$. Using the spherical degree value of refined salt particle is 0.71 ($\phi = 0.71$) and other parameters were taken from the **Table 5** which described in Bui (2009). Then we use the Ergun equations to calculate the minimum fluidization velocity. It is recommended that Re_{mf} had no limit [13–15].

b. Calculation based on the correlation between (Re_{mf}), (Ar) and Kozeny-Carman

In these two calculation methods, the parameters in the calculations are taken from the empirical results according to **Table 5**.

c. Determination of the (V_{mf}) value according to Wen and Yu

According to Wen and Yu methods, the bed voidage of the refined salt particle layer at minimum fluidization state (ε_{mf}) was unknown, but we had

Determination of the minimum fluidization velocity according to theoretical and empirical methods											
No	d_h (mm)	Minimum fluidization velocity									
		Ergun	Re andAr	Ko and Ca	Wen and Yu	Gedar	Go	Todes	Leva	Kunii and Levenspiel	Empirical methods
1	1.65	1.327	0.939	3.149	0.5414	0.45	1.1223	1.122	2.362	3.149	0.8
2	1.35	1.094	0.736	2.108	0.4002	0.331	0.9022	0.902	1.581	2.108	0.6
3	1.05	0.826	0.518	1.275	0.2625	0.219	0.6653	0.665	0.956	1.275	0.58
7	**0.953**	**0.731**	**0.445**	**1.05**	**0.2209**	**0.186**	**0.586**	**0.586**	**0.788**	**1.05**	**0.55**
4	0.75	0.522	0.298	0.651	0.1416	0.123	0.4188	0.419	0.488	0.651	0.42
5	0.45	0.22	0.115	0.234	0.0524	0.05	0.1863	0.186	0.176	0.234	0.2
6	0.225	0.058	0.029	0.059	0.0132	0.014	0.0537	0.054	0.044	0.059	0.15

*The bold values in **Table 4** refers to the common average particle size for commercial refined salt in the Vietnamese market.*

Table 4.
Theoretical calculation results of minimum fluidization velocity for refined salt particles from equations and empirical correlation formulas of published authors, which compare to the experimental results of the physical model of author.

Technical parameter	Symbol	Unit	Value	Ref.
Refined salt particle diameter	d_p	mm	1.65; 1.35; 1.05; 0.75; 0.45; 0.225	[13, 14] [15]
Average diameter of particle	d_m	mm	0.953	
Static bed voidage	ε_0		0.51	
Bed voidage in minimum fluidization velocity	ε_{mf}		0.56	
Particle density	ρ_p	kg/m^3	2138	
Bulk density	ρ_b	kg/m^3	982	
Spherical degree of particle	ϕ		0.71	
Gas density (at 160°C)	ρ_f	kg/m^3	0.815	
Dynamic viscosity (at 160°C)	μ_f	kg/m.s	2.45×10^{-5}	
Fixed refined salt particle bed height	H_0	m	30	

Note: The particle diameter d = 0.953 is the average diameter of the salt particles.

Table 5.
Physical parameters of refined salt grains and physical air.

the value of spherical degree particle from experiments (see **Table 5**). We put the value of spherical degree into Eq. (5) or Eq. (16) and found out the value of bed voidage (ε_{mf}) from which the velocity value of gas passing through the minimum fluidization particle layer (V_{mf}) was calculated.

d. Determination of V_{mf} according to Groshko-Todes

In this method, we also used the results of physical parameters of refined salt particles and air supplied to the dryer from **Table 5** to calculate Ar number. The Re_{mf} is determined by using Eq. 2 and the obtained result was multiplied by the error coefficient k = 1.2 and it was considered as the result of calculation of Re_{mf} for non-spherical salt particles (described by Lebedev, [21]). In addition, in Todes method there was another calculation by using Eq.(29) based on the available results of the minimum fluidization velocity (ε_{mf} = 0.4). We put this value into Eq. (15) and we found out the spherical degree of salt particles according to correlation given by Wen and Yu. Then we put this value into Eq. (12) to determine Ar number. By replacing Eq. (29) with the value of Ar number, we found out Re_{mf}, from which we could calculate V_{mf} value by using Eq. (10).

e. Determination of V_{mf} by the formula of Beayens-Geldart, Goroshko, Leva, and Kunii-Levenspiel

The minimum fluidization bed velocity (V_{mf}) was determined by using theoretical calculation of formulas of Beayens-Geldart, Goroshko, Leva, and Kunii-Levenspiel with the available physical parameters of refined salt particles and gas given in **Table 5** [18, 19, 22, 25, 26]. **Table 4** shows the calculated results from the formulas of authors published last time.

3.1.2 Primary conditions for determining V_{mf} by experimental method

A model in **Figure 3** and the other of fluidized bed dryer in **Figure 4** was designed by authors to define the minimum fluidization velocity of refined salt

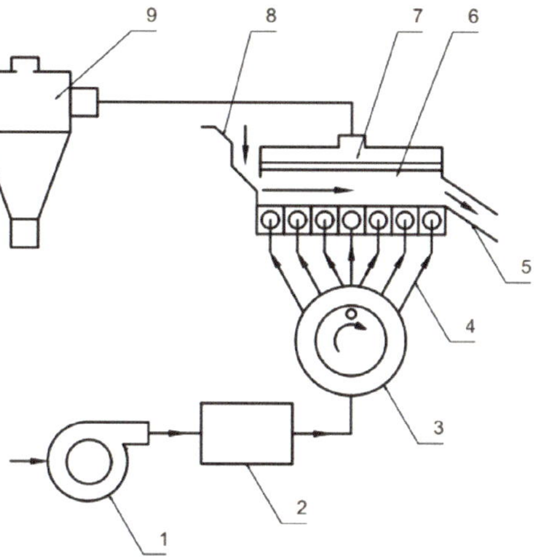

Figure 4.
The model of continuous fluidized bed dryer used in experiments. 1. Air fan; 2. heating chamber; 3. air supplier; 4. air duct; 5. product outlet; 6. drying chamber; 7. dust separation chamber; 8. inlet feeder; 9. cyclone dust collector.

particles in experiment. The dryer model was designed with its capacity of 48 kg/hour, the height of salt particle layer at the static bed was 30 mm ($H_0 \geq 30$ mm). In the experiments, the authors determined the minimum fluidization velocity (V_{mf}) for refined salt particles with diameter 1.65, 1.35, 1.05, 0.9, 0.65, 0.4 and 0.3 mm. The experimental results of determination of the minimum fluidization velocity of the particle layers with the different particle sizes are shown in **Table 4**. Besides, these experimental minimum fluidization velocity values were also compared with results of theoretical models that were published by authors presented in the methodology part above (**Figure 5**).

3.2 Discussions

- The obtained values of minimum fluidization velocity (V_{mf}) calculated by the Ergun equation and the correlation between Re_{mf} number and Ar number for

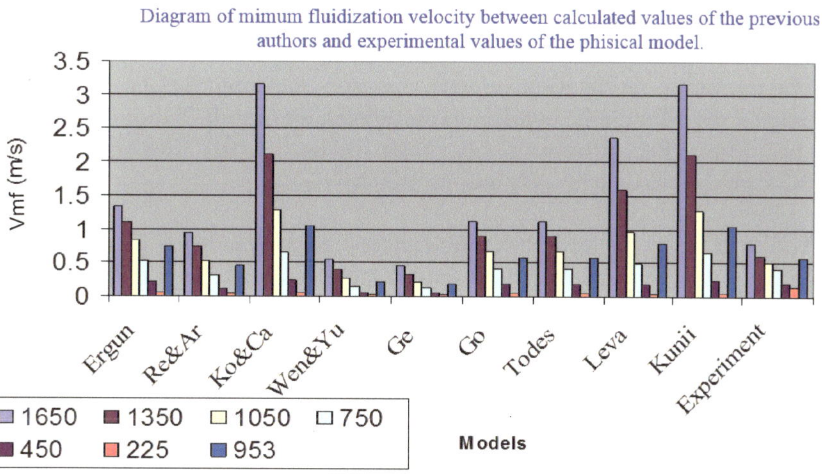

Figure 5.
Comparison of the minimum fluidization air velocity between calculated values of published authors and experimental values of the model [14].

all particle sizes agreed well with the experimental values. The value of Re_{mf} number varies from 0.3 to 51.7. Particles with sizes d_p = 0.225; 0.45 and 0.75 had the tendency for laminar flow.

- The obtained values of minimum fluidization velocity (v_{mf}) determined by the Kozeny-Carman and Kunii equations for particles with diameter greater than 1 mm ($d_p > 1$ mm) were much larger than experimental values ($V_{tmf} >> V_{emf}$). While, with particles having diameter smaller than 1 mm ($d_p < 1$ mm), the result of calculation was nearly equal to the experimental values. Notably, the void fraction value of the minimum fluidization state from 0.4 to 0.5 (ε_{mf} = 0.4–0.5) the calculation result matches the experimental value.

- By using the correlation of Wen and Yu to calculate void fraction at (ε_{mf}) knowing the spherical properties of particle, we obtained values that were much smaller than the experimental value. The Re_{mf} number varied from 0.07 to 21.1.

- When using the correlation between Re_{mf} and Ar, the obtained values of minimum fluidization velocity fit quite well to experimental results. Reynolds values vary from 0.16 to 36.6.

- The minimum fluidization velocity that was calculated according the correlation between Re_{mf} and Ar number of Beayens and Geldart (Eq. (19)) gave reasonable results.

- The obtained values by using the Goroshko and Todes formula in Eq. (28) were nearly equal to the experimental values.

- The calculated value of V_{mf} according to Beayens and Geldart was the lowest in comparison to other methods.

- The difference between the values of Re_{mf} calculated according to Goroshko and Ergun was lower than 10–20% for particles that lie in the range of Re_{mf} number from 0.28 to 43.7 (Re_{mf} = 0.28–43.7). The obtained values of velocity value (V_{mf}) for particles with diameter smaller than 0.9 mm ($d_p < 0.9$ mm) were closer to the experimental value in comparison with particles with diameter larger than 0.9 mm ($d_p > 0.9$ mm).

- The minimum fluidization velocity that was calculated by Leva formula was only suitable for particles with diameter smaller than 0.75 mm ($d_p < 0.75$ mm) and value of Re_{mf} number smaller than 10 ($Re_{mf} < 10$). The regime of air flow through the particle layer is laminar flow.

- The calculation method of the minimum fluidization velocity according to the Kunii and Levenspiel equations was suitable for particle with diameter d_p = (0.225; 0.45; 0.75 mm) and the results were appropriate under the conditions $Re_{mf} < 20$, and the calculation result was close to the experimental value.

3.3 Determination of air velocity through the particle layer at the optimum fluidization regime (V_{of})

The optimal fluidization velocity V_{of} was in the region from A to C (**Figure 2**). It meets the conditions:

$$V_{mf} < V_{hf} < V_{cf} \tag{38}$$

The air superficial velocity value from A to C (**Figure 2**) was determined by the two standard equations as follows:

The air velocity through the solid particle layer at the optimum fluidization regime (V_{of}) was calculated according to Fedorov standard (Fe) by Eq. (39) (as cited in Lebedev [21]), with refined salt particles and drying air parameters taken from **Table 5** (t_f = 160°C) [13, 15].

$$Fe = \phi d_p \left(\frac{4g\left(\rho_p - \rho_f\right)\rho_f}{3\mu_f^2} \right)^{1/3}$$
$$= 0.71 \times 0.956 \times 10^{-3}. \left(\frac{4 \times 9.8(2138 - 0.815)0.815}{3.\left(2.45 \times 10^{-5}\right)^2} \right)^{1/3} = 32 \tag{39}$$

According to Ginzburg (1973), Re_{hf2} was calculated by Eq. (40) [20, 27].

$$Re_{hf2} = (0.19 \div 0.285)\ Fe^{1.56} -> Re_{hf2} = 0.237 \times 32^{1.56} = 53 \tag{40}$$

The homogeneous fluidization velocity (V_{hf1}) was calculated by Eq. (41).

$$V_{hf2} = \frac{\mu_f . Re_{hf}}{\rho_f . d_p} = \frac{2.45 \times 10^{-5} \times 53}{0.815 \times 0.956 \times 10^{-3}} = 1.6 m/s \tag{41}$$

Reynolds value at homogeneous fluidization velocity (V_{hf1}) was measured according to Archimedes standard by Eq. (42) [20].

$$Re_{hf1} = (0.22 \div 0.33) \times Ar^{0.52} \tag{42}$$

where Ar was calculated by Eq. (12):

$$Ar = \frac{\rho_f\left(\rho_p - \rho_f\right)g\left(\phi d_p\right)^3}{\mu_f^2}$$

Substituting parameters of air and refined salt particles into the Eq. (12),

$$Ar = \frac{9.81 \times \left(0.71\left(956 \times 10^{-6}\right)\right)^3 \times 0.815 \times (2138 - 0.815)}{\left(2.45 \times 10^{-5}\right)^2} = 8818.4$$

The Ar value is 8818.4
Therefore, $Re_{hf1} = 0.275 \times (8818.4)^{0.52} = 66.917 = 31.53$.
The homogeneous fluidization velocity (V_{hf1}) was calculated by Eq. (43).

$$V_{hf1} = \frac{\mu_f . Re_{hf}}{\rho_f \phi d_p} = \frac{2.45 \times 10^{-5} \times 31.53}{0.815 \times 0.956 \times 10^{-3}} = 0.99\ m/s \tag{43}$$

Re-calculating the homogeneous fluidization velocity (V_{hf1}) by using the experimental equation Eq. (44) [27].

From **Table 4** for the specific case: V_{mf} = 0.55 m/s and particle diameter was 0.953 mm.

$$V_{hf1} = (2 \div 3) \times V_{mf} = (2 \div 3) \times 0.56 = (1.12 \div 1.68)\ m/s \tag{44}$$

Both V_{hf1} and V_{hf2} met the conditions of the Eq. (38). Selecting the optimum velocity V_{of}:

$$V_{of} = (V_{hf1} + V_{hf2})/2 = (0.99 + 1.66)/2 = 1.33\ m/s$$

Re-calculating the standard Reynolds number at reasonable fluidization state (Re_{of}) under the condition of optimum fluidization velocity (V_{of} = 1.33 m/s)

$$V_{of} = \frac{\mu_f . Re_{of}}{\rho_f . d_p} \tag{45}$$

or

$$Re_{of} = \frac{\phi d_p \rho_f V_{of}}{\mu_f} = \frac{0.71(0.956 \times 10^{-3})0.815 \times 1.33}{2.45 \times 10^{-5}} = 30$$

The value of optimum Reynolds number (Re_{of}) was 30 (Re_{of} = 30).

The void fraction of the particle layer at the reasonable fluidization state was determined by the Zabrodski formula (Eq. 46) (described in Lebedev, [21]).

$$\varepsilon_{hf1} = \left(\frac{18\, Re_{ohf} + 0.36\, Re_{ohf}^2}{Ar}\right)^{0.21} = \left(\frac{18 \times 30 + 0.36(30)^2}{8818.4}\right)^{0.21} = 0.61 \tag{46}$$

Re_{hf2} could be recalculated according to the correlation between Re_{hf2} and Ar Eq. (47).

$$Re_{hf2} = (0.22 \div 0.33) Ar^{0.52} = 0.275 \times (8818.4)^{0.52} = 31 \tag{47}$$

Substituting value of Re_{hf2} into Eq. (46), we re-calculated the homogeneous fluidization void fraction (ε_{hf}) by Eq. (48).

$$\varepsilon_{hf2} = \left(\frac{18\, Re_{hf} + 0.36\, Re_{hf}^2}{Ar}\right)^{0.21} = \left(\frac{18 \times 31 + 0.36(31)^2}{8818.4}\right)^{0.21} = 0.62 \tag{48}$$

These two calculation methods generated almost identical results.

3.4 Calculation of the critical air velocity flowing through the fluidization particle layer

In order to have a basis for determining the reasonable dimension of separating chamber of fluidized bed dryer (the chamber was located above the fluidization particle drying tank) and to limit removal of materials from the drying chamber, we defined the theoretical critical velocity (also called the final velocity).

From the calculation result of Re_{mf} of the minimum fluidization state ($Re_{mf} = 10.032$), this parameter of the air stream through the particle layer in the transition flow was in range $1 < Re_{mf} < 500$.

According to the equation of Haider and Levenspiel (described in Wen-ChingYang) [28, 29], the critical velocity was calculated by Eq. (49).

$$V_{cf} = \left[\frac{4\left(\rho_p - \rho_f\right)g}{3\rho_f C_D}\phi d_p\right]^{1/2} = \left[\frac{4(2138 - 0.815)9.81}{3 \times 0.815 \times C_D}0.71 \times 0.953\right]^{1/2} \quad (49)$$

With resistance coefficient (C_D) (described in Wen-ChingYang) [27, 28] given by

$$C_D = \frac{18}{\text{Re}^{3/5}} = \frac{18}{(10.032)^{3/5}} = 4.5 \quad (50)$$

So, the critical velocity at position C (**Figure 2**) in the specific case had the value:

$$\begin{aligned} V_{cf} &= \left[\frac{4\left(\rho_p - \rho_f\right)g}{3\rho_f C_D}\phi d_p\right]^{1/2} \\ &= \left[\frac{4(2138 - 0.815) \times 9.81}{3 \times 0.815 \times 4.5}\left(0.71 \times 956 \times 10^{-6}\right)\right]^{1/2} = 2.3 m/s \end{aligned} \quad (51)$$

In fact, during the drying process, to ensure the drying productivity and quality, the dryer operator must observe the fluidization particle layer and adjust the inlet doors of drying air capacity at appropriate the air velocity value in the range from A to C (**Figure 2**) and the correlation of velocity types $V_{mf} < V_{hf} < V_{cf}$.

We re-calculated the void fraction of particle layer with average particle diameter $d_p = \phi d_m = 0.953$ mm at the theoretical critical velocity in fluidization particle layer condition.

$$\text{Re} - \text{calculating } \text{Re}_{cf} : \text{Re}_{cf} = \frac{\phi d_p . \rho_f . V_{cf}}{\mu_f} \quad (52)$$

Substituting the above value into Eq. (52), we get:

$$\text{Re}_{cf} = \frac{0.71 \times \left(956 \times 10^{-6}\right) \times 0.815 \times 2.3}{2.4 \times 10^{-5}} = 0.53.$$

Using Eq. (46) to re-calculate the void fraction of particle layer at the complete fluidization state:

$$\varepsilon_{tf} = \left(\frac{18\,\text{Re}_{cf} + 0.36\,\text{Re}_{cf}^2}{Ar}\right)^{0.21} = \left(\frac{18 \times 53 + 0.36 \times (53)^2}{8818.4}\right)^{0.21} = 0.73$$

When the void fraction of particle layer was 1 ($\varepsilon = 1$), the fluidization particle layer turned to the transport regime in the air stream (called pneumatic transportation).

4. Conclusions

Most of the used correlations in the calculations and the formulas given by the authors, as mentioned above, were derived from the experiments with temperature close to the ambient temperature. So, when we use them in calculations in specific cases, they should consider the accuracy. The extrapolation should be used in the cases of the states at the temperature higher than the ambient temperature.

The mentioned theoretical calculations show the necessity for the determining of the minimum fluidization velocity of the solid particle layer with high accuracy. The sphericity of particle and void fraction of the particle layer were often not known, therefore it is required to get their values from the range of experimental variables. Empirically, the void fraction of the particles in the minimum fluidization layer at the ambient temperature is not the same as that in the increasing gas temperature.

The best method to determine the minimum fluidization velocity is to conduct the experiments. Firstly, we directly measured the pressure drop across the particle layer when the air velocity gradually decreased. Secondly, we built the graphs and read the results of the minimum fluidization velocity value.

However, if we were forced to find out the fluidization velocity without carrying out experiments to measure the pressure drop across the particle layer, the best way would be to determine the void fraction at the minimum fluidization velocity. Then we calculated the spherical property of the particle using the Ergun equations or correlation between Ar and Re_{mf} in Eqs. (10)–(12) to count out the minimum air velocity through the particle fluidization layer. This velocity value also had accuracy close to the experimental one.

The average particle diameter considered spherical degree of the particles of different sizes was 953 μm (d_m = 953 μm). This diameter represented the size of the particles in dry grinding technology with the hammer crusher. It is also in the common size distribution range of the combined washing-grinding hydraulic-separation technology in Vietnam's market. Besides, the particle diameter of 953 μm (d_p = 953 μm) is also used in calculating the value of all types of velocity, characterized for the medium particle size of the refined salt production technology in Vietnam.

We calculated the values of three characteristic velocity types of fluidized bed drying for particles with average size $d_m = d_p$ = 953 μm, including the minimum fluidization velocity V_{mf} = 0.55 m/s with void fraction ε_{mf} = 0.56; reasonable fluidization velocity V_{hf} = 1.33 m/s corresponding to the void fraction of particle layer ε_{hf} = 0.615; and the critical velocity of the air flow through the particle bulk V_{cf} = 2.3 m/s with the void fraction value of fluidization particle layer ε_{cf} = 0.73.

In fact, during the drying process, to ensure the drying productivity and quality, the dryer operator must observe the fluidization particle layer and adjust the inlet doors of drying air capacity at appropriate air velocity value in the range from A to C (**Figure 2**) to make sure the correlation of velocity types $V_{mf} < V_{hf} < V_{cf}$.

Nomenclature

A	cross sectional area of the bed
m^2	area for gas distribution
H	height of the bed, m
Ar	Archimedes number, dimensionless
H_0	initial bed height, m

ε_0	void fraction at static particle layer
H_{mf}	minimum fluidization bed height, m
ε_{mf}	void fraction at minimum fluidization particle layer
ΔP	pressure drop across the bed, N/m^2
ε_{hf}	void fraction at homogeneous fluidization particle layer
Re	Reynolds number
ε_{tf}	terminal void fraction
Re_{mf}	Reynolds number at the minimum fluidization velocity
Re_{hf}	Reynolds number at the homogeneous fluidization velocity
d_p	spherical particle diameter, m
ϕd_m	equivalent spherical mean diameters, m
Re_{cf}	Reynolds number at the fluidization terminal velocity
ϕ	sphericity degree, dimensionless
Re_{of}	Reynolds number at the optimum fluidization velocity
Fe	Fedorov standard, dimensionless
V	bed surface velocity or superficial velocity, m/s
g	acceleration due to gravity, m/s^2
V_{mf}	minimum fluidization velocity, m/s
ρ_p	solid particle density, kg/m^3
V_{tmf}	theoretical minimum fluidization velocity, m/s
ρ_b	particle bulk density, kg/m^3
V_{emf}	experimental minimum fluidization velocity, m/s
ρ_f	air density, kg/m^3
V_{hf}	homogeneous fluidization velocity, m/s
Kcc	Kozeny-Carman coefficient
V_{ohf}	optimum homogeneous fluidization velocity, m/s
m_p	mass of particles, kg
V_{cf}	terminal velocity or critical velocity, m/s
m_b	mass of particles bulk, kg
μ	air dynamic viscosity, kg/ms
S_s	specific surface area, cm^{-1}, cm^2/g
U	cubic volume of particle layer, m^3
W	weight of particle, mass, N

Author details

Bui Trung Thanh[1]* and Le Anh Duc[2]*

1 Industrial University of Ho Chi Minh City, Vietnam

2 Nong Lam University, Ho Chi Minh City, Vietnam

*Address all correspondence to:
buitrungthanh@iuh.edu.vn and leanhduc@hcmuaf.edu.vn

References

[1] Tavoulareas S. Fluidized-bed combustion technology. Annual Reviews. 1991;**16**:25-27

[2] Lewis WK, Gilliland ER, Bauer WC. Characteristics of fluidized particles. Industrial and Engineering Chemistry. 1949;**41**:1104-1117. DOI: 10.1021/ie50497a059

[3] Zahed AH, Zhu JX, Grace JR. Modelling and simulation of batch and continuous fluidized bed dryers. Drying Technology. 1995;**13**:1-28. DOI: 10.1080/07373939508916940

[4] Han JW, Keum DH, Kim W, Duc LA, Cho SH, Kim H. Circulating concurrentflow drying simulation of rapeseed. Journal of Biosystems Engineering. 2010;**35**(6):401-407. DOI: 10.5307/JBE.2010.35.6.401

[5] Hong SJ, Duc LA, Han JW, Kim H, Kim YH, Keum DH. Physical properties of rapeseed (II). Journal of Biosystems Engineering. 2008;**33**(3):173-178. DOI: 10.5307/JBE.2008.33.3.173

[6] Duc LA, Han JW. The effects of drying conditions on the germination properties of rapeseed. Journal of Biosystems Engineering. 2009;**34**(1): 30-36. DOI: 10.5307/JBE.2009.34.1.030

[7] Abrahamsen AR, Gedart D. Behaviour of gas-fluidized beds of fine powders part I. Homogeneous expansion. Powder Technology. 1980; **26**:47-55. DOI: 10.1016/0032-5910(80) 85006-6

[8] Geldart D. The effect of particle size and size distribution on the behaviour of gas-fluidized beds. Powder Technology. 1972;**6**:201-215

[9] Geldart D. Types of gas fluidization. Powder Technology. 1973;7:285-292. DOI: 10.1016/0032-5910(73)80037-3

[10] Howard JR. Fluidized Bed Technology: Principles and Application. Publisher Taylor and Francis Group; 1989. p. 214

[11] Rhodes M. Fluidization of particles by fluids. In: Educational Resources for Particles Technology. Melbourne, Australia: Monash University; 2001. pp. 1-39

[12] Ergun S. Fluid flow through packed columns. Chemical Engineering Progress. 1952;**48**:9-94

[13] Bui TT. Determination on geometrical parameters of refined salt particle applying of fluidized particle lay drying. Journal of Heat Science and Technology. 2009;**86**:10-13

[14] Bui TT. Determination on hydrodynamic parameters in fine salt drying in a model of fluidized bed dryer. Journal of Heat Science and Technology. 2009;**90**:13-17

[15] Bui TT. Researching and determining on basic physical parameters of refined salt particles to apply in the calculation and designing of continuous fluidized bed. Journal of Vietnam Mechanical Engineering. 2009; **146**:8-31 and 48

[16] Yates JG. Fundamentals of Fluidized-Bed Chemical Processes. 1st ed. Oxford: Butterworth-Heinemann; 1983. p. 236

[17] Wen CY, Yu YH. A generalized method for predicting the minimum fluidized velocity. AICHE Journal. 1966; **12**:610-612

[18] Baeyens J, Geldart D. Predictive calculations of flow parameters in gas fluidized bed and fluidization behavior of various powders. In: Proceedings of

the International Symposiu on Fluidization and its Applications; 1973

[19] Goroshko VD, Rozenbaum RB, Todes OM. Approximate laws of fluidized bed hydraulics and restrained fall. Izvestiya Vysshikh Uchebnykh Zavedenii, Neft i Gaz. 1958;**1**:125-131

[20] Tran VP. Calculation and Designing on the Drying System. Hanoi: Education Publishing House; 2002

[21] Lebedev PD. Calculation and Design of Drying Equipmet. Moscow, Leningrad: Gosenergoizdat; 1963. p. 320 (in Russian)

[22] Leva M. Fluidization. New York: McGraw-Hill; 1959. p. 281

[23] Carman PC. Fluid flow through granular beds. Transactions, Institution of Chemical Engineers. 1937;**15**:150-166

[24] McCabe WE, Smith JC, Harriott P. Unit Operations of Chemical Engineering. 6th ed. New York: McGraw Hill; 2001. p. 1132

[25] Kunii D, Levenspiel O. Various kinds of contacting of a batch of solids by fluid. In: Fluidized Engineering. Huntington, NY: Robert E. Krieger Publishing Co.; 1977. pp. 24-56

[26] Kunii D, Levenspiel O. Fluidization engineering. 2nd ed. Boston: Butterworth-Heinemann; 1991. p. 49

[27] Ginzburg AS. Theoretical and Technical Basis of Drying Food Products. Moscow: Pishchevaya Promyshlennost; 1973. p. 528 (in Russian)

[28] Yang WC. Particle characterization and dynamics. In: Yang WC, editor. Handbook of Fluidization and Fluid-Particle System. New York: Marcel Dekker; 2003. pp. 1-29

[29] Yang WC. Flow through fixed beds. In: Yang W-C, editor. Handbook of Fluidization and Fluid-Particle System. New York: Marcel Dekker; 2003. pp. 29-53

Section 3

Novel Drying Technologies

Chapter 6

The Study of Fabric Drying Using Direct-Contact Ultrasonic Vibration

Chang Peng and Saeed Moghaddam

Abstract

Our existing cloth drying technology is an energy-intensive process, which generally involves blowing hot air across tumbling wet fabrics to facilitate evaporation and moisture removal. To address the relatively low energy efficiency of existing cloth drying techniques, in this chapter, a totally new cloth drying technology is introduced, which uses high frequency ultrasonic vibrations generated by piezoelectric transducer instead of thermal heating to extract moisture in cloth as a cold mist, dramatically reducing drying time and energy consumption. The physical mechanism of ultrasonic fabric drying process in direct-contact mode is first studied. A novel ultrasonic transducer driving method, in which the power supply to the transducer is regulated by a binary modulating signal, is then developed for use in direct-contact ultrasonic drying of fabrics. A demonstration unit is finally fabricated to show the efficacy of the process and its energy saving compared to thermal drying process.

Keywords: fabric drying, energy efficiency, ultrasonic vibration, ultrasonic transducer, physical mechanism

1. Introduction

Clothes dryers offer a rapid means to dry laundry in households but consume substantial residential electricity. Clothes dryers are the second largest source of residential energy use in a household following water heater [1]. Every year the operation of clothes dryers in the United States consumes as much electricity as does the entire state of Massachusetts (60 billion kWh per year) [2]. The demand of clothes dryers in U.S. has increased by nearly 6% over the last decade. Until 2009, 80% of U.S. households had a clothes dryer and 80% of them are electric clothes dryers, the remaining being gas dryers [3]. Although a set of technologies such as air-vented dryers and heat pump dryers have been introduced to the market, the energy efficiency of clothes dryers has not been improved significantly over the last 20 years.

In the current clothes dryer architecture, the clothes dry by passing hot air over the wet fabric. The state of the art for clothes drying technology involves mechanical (centrifugal) water extraction in the washing machine followed by thermal drying in a dryer. The moisture initially extracted by the washing machine reduces the remaining moisture in the clothes to ~50% of the dry weight [4]. The clothes are then placed in a dryer drum, where the remaining moisture is removed by circulating heated air through them. Thermal water removal from clothes in all existing

Dryer type	Average drying time (min)	Energy factor (lb/kWh)
Electric standard (≥4.4 ft^3 capacity) [3]	40	3.73
Electric compact (120 V, ≤4.4 ft^3 capacity) [3]	40	3.61
Electric compact (240 V, ≤4.4 ft^3 capacity) [3]	35	3.27
Gas standard [3]	35	3.30
Heat pump [2]	82	7.6

Table 1.
Average clothes drying time and energy factor (EF) values for different types of commercialized clothes dryers.

dryer technologies [3, 5–10] (including heat, heat pump, vacuum, or microwave drying) requires a heat source to provide a latent heat of evaporation of about 2.5 MJ/kg water. The heat source for a dryer can be electric resistance, natural gas and/or an electric heat pump.

The metrics to measure the drying efficiency of a clothes dryer is called energy factor (EF), which commonly has three different types of definitions, depending on how the drying effect is qualified: (1) the mass of moisture removed per energy consumption; (2) the mass of cloth dried per energy consumption; or (3) the latent heat of moisture removed per energy consumption. In the United States, the Department of Energy standard of EF measures the mass of cloth (saturated to certain water content) that can be dried per unit of electricity consumed (lb/kWh) [3]. For gas dryers, it measures the pounds of clothes being dried per equivalent kilowatt-hour of natural gas consumed. Here, we focus on EF (lb/kWh) which is commonly used in the United States. **Table 1** summarizes the EF values for the typical types of existing clothes dryers.

Recently, for the first time, we have introduced a totally new cloth drying technology and are collaborating with Oak Ridge National Laboratory (ORNL) researchers to develop the world's first direct-contact ultrasonic cloth dryer that could potentially change the concept of residential heat-drying technology [11–15]. This novel approach uses high-frequency mechanical vibrations generated by piezoelectric ceramics instead of heat to extract moisture as a cold mist, dramatically reducing drying time and energy use. On the basis of inverse piezoelectric effect, when an electric field is applied in the polarization direction of piezoelectric ceramics, the ceramics will produce mechanical deformation in a certain direction. When a high frequency electric field is applied, the piezoelectric transducer generates acoustic vibration. If the oscillation frequency is larger than 20 kHz, it generates ultrasonic waves that propagate through its surrounding media [12, 16]. Coming to the studies reported on ultrasonic cloth drying per se, to our knowledge, Khmelev et al. [17, 18] is the only paper on this topic. However, a non-contact technique was used in their study, making our study unique and the first of its kind.

This study will address the feasibility of adopting the ultrasonic technique for cloth drying with the objective of achieving lower power consumption and process duration compared to the existing energy-intensive, thermal-based clothes dryers.

2. Experimental setup and method

A schematic of the experimental setup for studying both the water droplet atomization and the fabric drying process is shown in **Figure 1**. The setup mainly consists of six components: an ultrasonic transducer, a function generator, a power

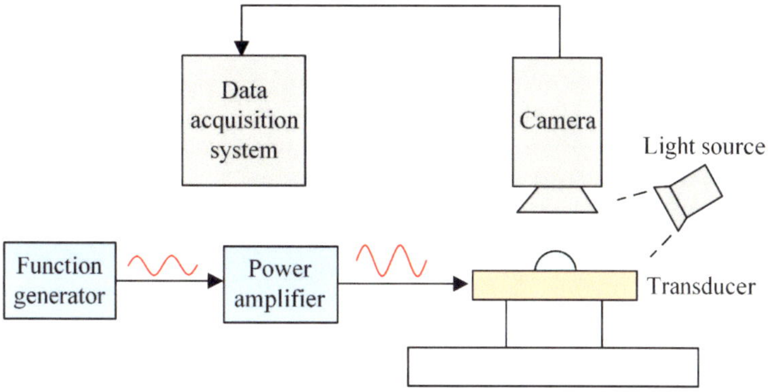

Figure 1.
Schematic representation of the experimental setup for droplet atomization and fabric drying studies [11].

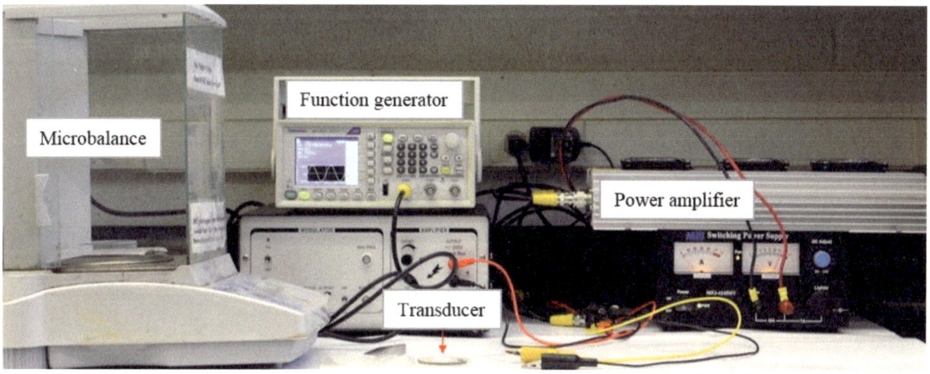

Figure 2.
Experimental setup for fabric drying study [11].

Transducer	Working diameter (mm)	Resonant frequency (kHz)	Type
A	20	1600	Piezo crystal
B	50	1060	Piezo crystal
C	4.2	135	Metal mesh

Table 2.
Parameters of transducers tested in this study [11].

amplifier, a high-speed camera, an infrared (IR) camera and a microbalance with a resolution of 0.1 mg for measuring the fabric weights. An image of the experimental setup is provided in **Figure 2**.

The detailed characteristics of the transducers tested in this study are summarized in **Table 2**. Transducer A and B are traditional piezoelectric transducers, the working surfaces of which consist of a piezoelectric ceramic, as shown in **Figure 3a–b**. Transducer C is a metal mesh-based transducer, which consists of a thin metal mesh sandwiched between two piezoelectric rings (**Figure 3c**) that vibrate the mesh. If a water droplet is in direct contact with the mesh surface, the metal mesh–based transducers allow water to pass through the mesh.

For the water droplet atomization study, a water droplet of a specified volume was deployed at the center of the transducer by using a micro-pipette. The atomization process was monitored using the high-speed camera at a frame rate of 3600

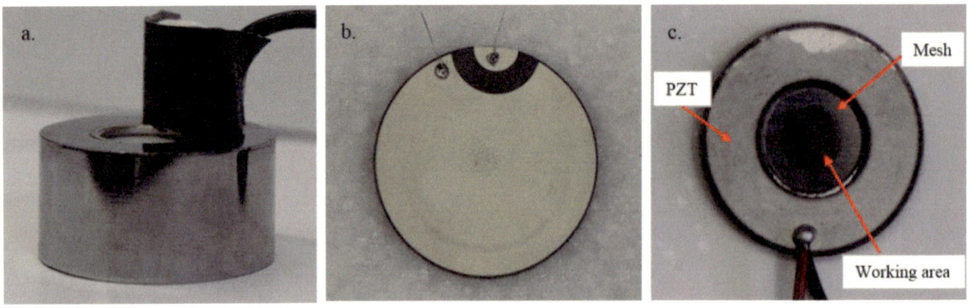

Figure 3.
Images of the transducers used in this study: (a) transducer A, (b) transducer B, and (c) transducer C [11].

Figure 4.
Scanning electron microscope images of the two DOE fabrics [11].

frames per second. The camera line of view was normal to the vibrating surface, thus providing top-view images of the atomizing droplets (**Figure 1**) [11]. During the water droplet atomization process, the IR camera was also used to monitor the droplet and transducer temperature changes. It was mounted in the same way as the high-speed camera. Trials were performed with the following values of the initial droplet volume: 10, 50, and 100 μL.

In the studies, two kinds of US Department of Energy (DOE) standard fabrics (hereinafter called F-1 and F-2) were tested. The SEM images of the fabrics are shown in **Figure 4**. As can be seen from **Figure 4**, F-1 has larger major pore sizes than F-2. Details on the contents, thicknesses and densities for these two fabrics can be found in Ref. [11].

In the studies, the microbalance, AL104, with an error of 0.1 mg was used to measure the weight of the saturated and unsaturated fabric samples to determine the volume of water remaining in the fabric. Prior to the drying test, the dry fabric was placed on the transducer and the weight of this system was measured on the microbalance. After this measurement, the fabric was wet with a known volume of water. The transducer was then actuated to begin the drying process. After 5 s elapsed, the transducer was turned off and the weight of the transducer-fabric system was measured. Immediately after this measurement, the fabric was completely dried by heating it to 80°C. The fabric was then rewet and the weight of the transducer-fabric system was measured after actuating the transducer for 10 s. The fabric was then completely dried. In this manner, the weight of an initially saturated transducer-fabric system was measured after actuating the transducer in increments of 5–10 s, with the fabric dried after each measurement. For each time interval, the tests were repeated for at least three times. The measurements were recorded until the weight of the transducer-fabric system was less than 5% of the initially saturated system. This method of determining the drying rate of fabrics was repeated for the two fabrics on the three transducers.

3. Ultrasonic atomization of water droplet

Figure 5 illustrates the high-speed images recorded during the atomization of a 10 μL water droplet on transducer A. A qualitative analysis of these images can provide clues to the kinetics of the atomization process. In the initial stage after the transducer was turned on, the oscillations of the transducer surface were transmitted to the surface of the droplet resulting in the formation of capillary waves (**Figure 5b**). As time progressed, the capillary waves grew to form surface ligaments (**Figure 5c**), followed by the atomization of droplet (**Figure 5d–j**), which the droplet broke up via ejection of secondary droplets. The atomization of the droplet exhibited a very chaotic dynamic. In addition to the production of fine mist from the droplet surface, the droplet was observed to spread and deform during the process. [11].

The observed kinematics of droplet atomization can be explained by analyzing the relative magnitudes of the oscillation force (F_o) and surface tension force (F_s) influencing the process [11]. Assuming that the droplet forms a half-sphere on the transducer surface (**Figure 6**), the scale of F_o applied on the water droplet (droplet mass, m_{drop}, and density, ρ) can be presented as,

$$F_0 \sim m_{drop} A\,\omega^2 \sim \frac{1}{12}\pi d^3 \rho A\,\omega^2 \tag{1}$$

where A and ω are the vibration amplitude and angular frequency ($\omega = 2\pi f$), respectively. On the other hand, the scale of the surface tension force (F_s) experienced by the droplet surface can be presented as [19],

$$F_s = \sigma \pi d cos\theta \tag{2}$$

where σ is the liquid surface tension which is temperature (T) dependent; d is the droplet diameter which varies with time (t); and θ represents the contact angle formed at the liquid–solid interface.

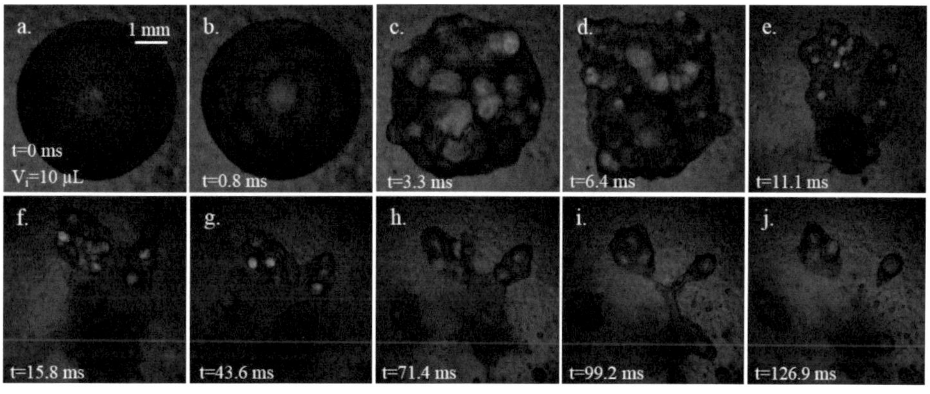

Figure 5.
Different regimes of droplet atomization process on transducer A: (a) unperturbed droplet, (b) capillary wave regime, (c) ligament regime, and (d)–(j) catastrophic drop breakup and atomization [11].

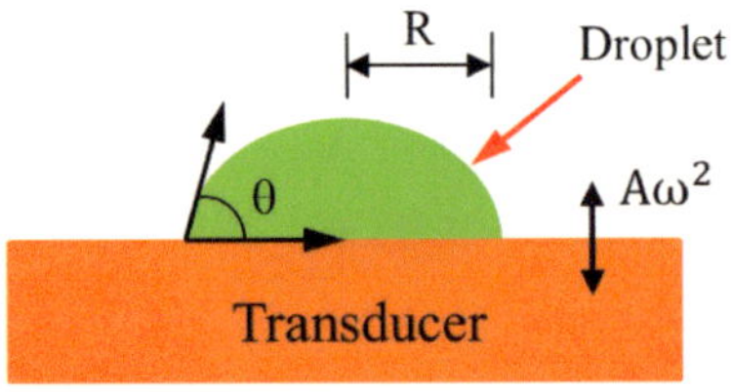

Figure 6.
Schematic of a water droplet on a vibrated transducer surface.

The relative magnitude of F_o and F_s can be expressed as,

$$R \sim \frac{F_0}{F_s} \sim \frac{d^2 \rho A \omega^2}{12\sigma cos\theta} \tag{3}$$

For a 10 μL water droplet on transducer A (**Figure 5a**), $F_o > F_s$ (i.e. $R > 1$) when the transducer was turned on. This caused the liquid–vapor interface to rupture (leading to atomization), reducing the droplet radius (r) and R. The atomization continued with time until $F_s > F_o$ (at a critical volume), at which point, the atomization process ceased.

During the ultrasonic droplet atomization process, the droplet temperature was also changed due to the viscous dissipation of continuous vibration. And as illustrated in Eq. 3, the surface tension of the droplet (σ) is temperature (T) dependent. Higher the droplet temperature, lower the surface tension will be. The lower surface tension will result in the lower surface tension force (F_s) (Eq. 3). In the following section, the temperature changes on the two types of transducers, piezoelectric crystal transducer and metal mesh transducer, were studied during the droplet atomization process.

Three different volumes of water, 10, 50, and 100 μL, were tested on transducers A, B and C. The thermal responses of these droplets were recorded using the IR camera and the maximum liquid temperature was extracted from this data. **Figure 7** presents the temporal variations in the maximum temperatures of these droplets on transducers. The x-axis represents a non-dimensional time (t^*) defined as

$$t^* = \frac{t}{t_0} \tag{4}$$

where t_0 represents the total duration of time that was recorded using the IR camera (**Tables 3** and **4**).

As can be seen from **Figure 7a**, the temperature for 10 μL droplet varied little immediately after the transducer A was actuated. The temperature went up to

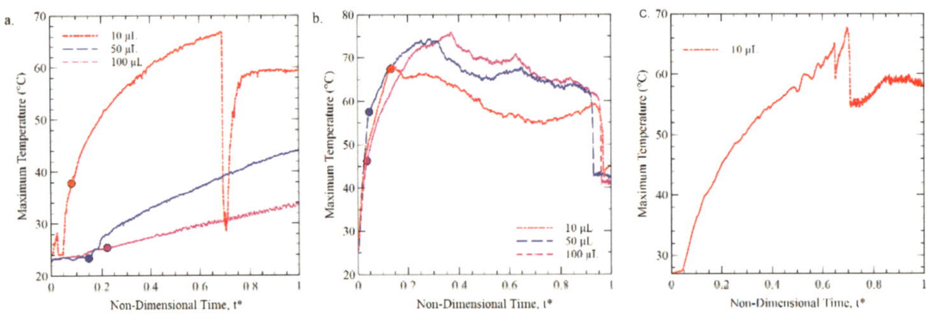

Figure 7.
The maximum temperature of liquid on the surface of (a) transducer A, (b) transducer B and (c) transducer C, extracted from the IR images recorded during liquid atomization [11].

Initial volume (μL)	Process time, t_o (s)		
	Transducer A	Transducer B	Transducer C
10	3.29	157.33	31.83
50	3.23	248.67	—
100	2.43	265.00	—

Table 3.
Summary of the total process time for different droplet volumes on transducers A, B and C [11].

Droplet volume (μL)	Non-dimensional time (t^*)	Threshold temperature (°C)
10	0.13	67.4
50	0.05	57.5
100	0.04	46.3

Table 4.
Threshold temperature and non-dimensional time at which atomization was induced on transducer B [11].

the peak value and then went down sharply. After that process, the temperature increased again and then kept stable. However, for 50 and 100 μL droplet, the temperature just showed the increase in the whole process in **Figure 7a**. And the maximum temperature was much smaller than that of the 10 μL droplet on transducer A. As discussed before, $F_o > F_s$, the droplets could be burst out on transducer A immediately after the transducer was actuated. Meanwhile, the temperature on transducer A kept stable in this tiny time interval as shown in **Figure 7a**. After that the temperature increased due to the viscous dissipation.

For the transducer B in **Figure 7b**, $F_o < F_s$ when the transducer was turned on, the water droplet just vibrated continuously on the transducer B. And because of viscous energy dissipation, the temperature of the droplet went up quickly during the droplet vibrations as shown in **Figure 7b**. When the temperature on transducer B reached a peak value, the droplet on transducer B began to break up. As discussed before, this was because the surface tension of water (σ) decreased with the increase of temperature (T) and $F_s < F_o$. Meanwhile, when the secondary droplets were burst out, the temperature on transducer B decreased continuously.

For the droplet 10 μL on the transducer C, $F_o > F_s$ when the transducer was turned on, large amounts of droplets were jetted out. At this small time interval the temperature on transducer C kept stable, which was similar to the thermal response of the 10 μL droplet on transducer A.

4. Ultrasonic drying behavior of fabric under continuous driving

The baseline dewatering time of fabrics was determined by thermally drying two different saturated fabrics (F-1 and F-2) with the size 4 × 4 mm^2 at 80°C (**Figure 8**). The saturation volumes for the 4 × 4 mm^2 fabrics, F-1 and F-2, were ~6.45 and ~8.55 μL respectively. It was observed that the water content in a given fabric (via thermal evaporation) varied linearly with time (**Figure 8**), since it was directly related to the amount of water stored in the fabric.

Figure 9 demonstrates the drying characteristics of saturated fabrics F-1 and F-2 on transducers A, B, and C. The fabric size was selected such that the sample area did not exceed the working area of the transducers. It was seen that water

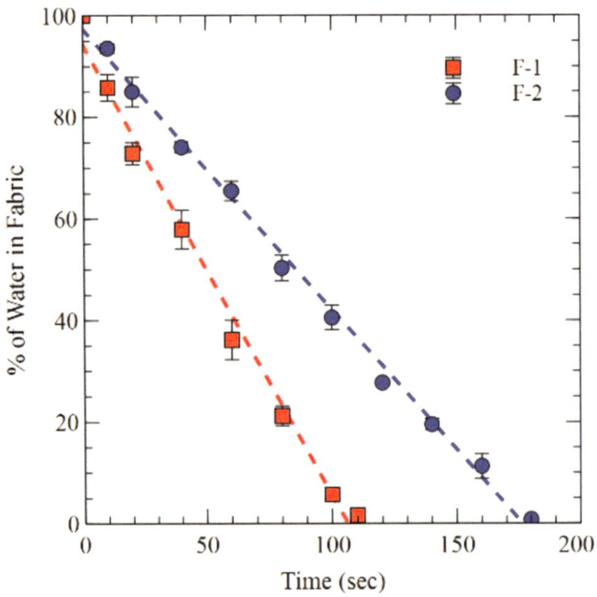

Figure 8.
Thermal drying behavior of two different fabrics [11].

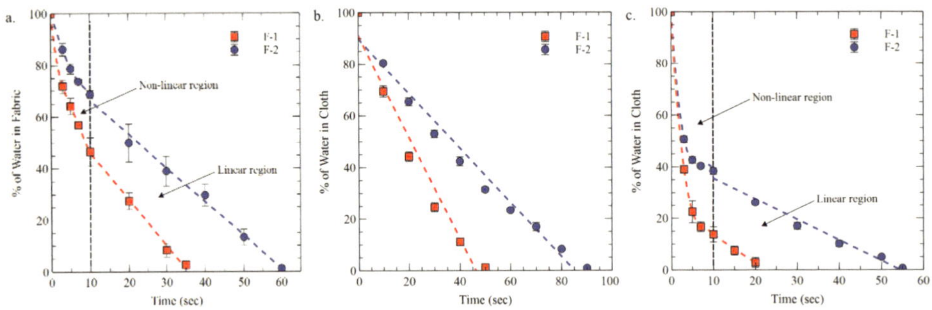

Figure 9.
Drying behavior of different types of fabric on (a) transducer A, (b) transducer B, and (c) transducer C [11].

shedding via ultrasonic excitation dewatered fabrics quicker than thermal drying (**Figure 8**), with up to an 82% reduction in drying time for fabric F-1 on transducer C. The data in **Figure 9** show that the drying curve for each fabric on transducers A and C consist of two distinct regions (indicated by the arrows in **Figure 9**):

- An initial region where the water content in the fabric decreases non-linearly with time;
- A second region where the water content in the fabric decreases linearly with time.

These two regions of the drying curve arise due to the impact of the two major forces involved in the water shedding process—the oscillation force (F_o) and the surface tension force (F_s). There are two kinds of pores in the fabrics: the large pores between the fabric threads and the small pores within the fabric threads (**Figure 4**). Due to the pore size distribution in the fabric, the surface tension force (F_s) in the fabric spans a range of values—large pores in the fabric generate low F_s while small pores generate large F_s. At $t^* = 0$, the oscillation force generated by the transducer surface (F_o) was sufficiently large to overcome the surface tension forces (F_s) in the larger pores of the fabric, resulting in rapid shedding of water from these pores. The shedding of a large fraction of water via ultrasonic excitation resulted in a non-linear decrease in the fabric water content. The fraction of water shedding in this region was dictated by the fraction of pores in the fabric with $F_o > F_s$, with the amount of water shedding increasing as the number of such pores increase. Based on the data presented in **Figure 9**, it can be deduced that fabric F-1 consist of a larger fraction of pores with $F_o > F_s$ than fabric F-2, resulting in a greater decrease in water content during the initial non-linear phase. This conclusion can also get the evidence from **Figure 4**, which fabric F-2 shows denser structure and smaller pores compared with fabric F-1. At the end of the non-linear phase, the only pores containing liquid were those with $F_s > F_o$. The continued excitation of the transducer surface caused the temperature of liquid trapped in these pores to rise via viscous dissipation without shedding water from the fabric. This response was similar to the thermal response of droplets observed when $F_s > F_o$. The increase in temperature of the liquid (and the fabric) induced thermal evaporation from the fabric, which caused a linear decrease in the fabric water content (similar to the trend presented in **Figure 8**).

Here, the kinematics of fabric drying on transducer B were similar to those of droplet atomization, where $F_s > F_o$ at $t^* = 0$. Therefore, the primary response of the liquid upon excitation was thermal, and the fabric temperature rapidly increased due to viscous dissipation. However, since the surface tension force (F_s) in the fabric was much greater than that of the 10 μL droplet, the increase in temperature did not cause F_s to decrease below F_o. Therefore, ultrasonic drying was not induced on this transducer, resulting in the absence of a non-linear drying region. Compared with the thermal drying curves for different fabrics (**Figure 8**), it also indicates that the water shedding on transducer B was primarily driven by thermal evaporation. As in the case of thermal drying, the drying time of the F-1 fabric was shorter than that of the F-2 fabric.

In addition, the drying data suggest that a larger volume of water was shed from the fabric during the nonlinear regime on transducer C than on transducer A and, overall, fabrics dried more quickly on transducer C than on transducer A and B.

5. Ultrasonic drying behavior of fabric under burst-mode driving

A simple analysis of the dynamics of the contact between the fabric and ultrasonic transducer during the fabric drying process illustrates that the transducer does not need to be constantly on to obtain the same drying effect. As the transducer vibrates, it pushes the fabric upward. The fabric loses contact with the surface of the transducer. After some time, the fabric falls back onto the transducer due to the gravity. During the time when there is no contact between the fabric and the transducer, power is consumed with no contribution to the fabric drying [12]. Therefore, power can be supplied to the transducer in bursts rather than continuously without affecting the drying process, which can increase the energy efficiency.

The drying behavior of fabric F-1 at different duty cycles for a given modulating frequency on transducer C is presented in **Figure 10**. The fabric size was fixed at 4×4 mm^2 and the initial water mass was 10 mg. The modulating frequency in **Figure 10a–b** is 100 and 1000 Hz, respectively. The drying behavior of the fabric with the modulating signal added keeps the same with the drying behavior of the fabric without modulating signal (represented by "No DC" in **Figure 10**), as illustrated in **Figure 10**. The total drying time increased from 20 to 30 s when the duty cycle was 20%. However, when the duty cycle was 40%, the drying time was 25 s and was 5 s less than the total drying time when the duty cycle was 20%, as

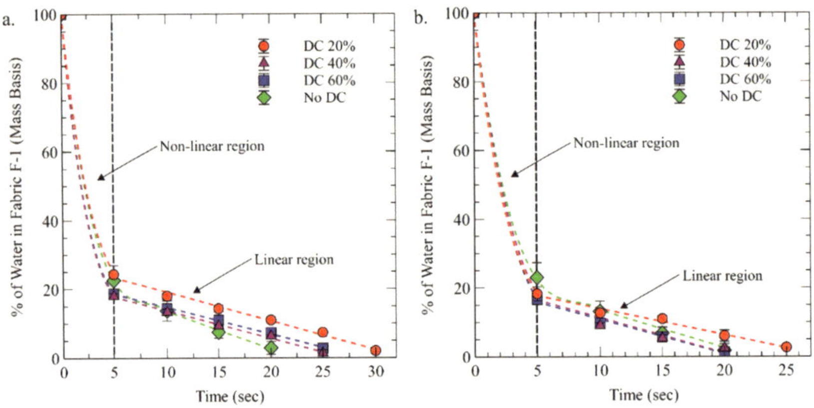

Figure 10.
Effect of duty cycle (DC) on the drying curves of fabric F-1 on transducer C. Modulating frequency: (a) 100 Hz and (b) 1000 Hz [12].

illustrated in **Figure 10(a)**. Increasing the duty cycle from 20 to 60% did not significantly affect the total drying time of the non-linear region of the drying curve. However, the slope of the linear region of the drying curve changed when the duty cycle was below 40%, resulting in an increase in the drying time.

When the transducer was turned on, because the ultrasonic excitation force of transducer C was much larger than the surface tension force of water in the large pores (i.e., $R > 1$), regardless of the duty cycle, the water accumulated in the large pores could be ejected quickly. However, water accumulated in the small pores experienced a larger surface tension force than the ultrasonic excitation force ($R < 1$). Therefore, the non-linear region of the drying behavior corresponded to water removal from the large pores while the linear region corresponded to drying in the small pores. In addition, in **Figure 10(a)**, it illustrates that adding the duty cycle did not change the nonlinear region of the drying process, but it changed the linear region of the drying process. This is also because the water embedded in the larger pores was easier to be atomized immediately after the transducer was actuated. As illustrated in **Figure 10(a)**, compared with the drying time in the nonlinear region, the drying time of the linear region was 3 times of that in the nonlinear region which there was no duty cycle added although the water percentage in the linear region was just 25% of the nonlinear region. For the duty cycle added, the linear region took even longer time compared with no duty cycle added. This is because the power in this situation was not continuously turned on and was determined by the duty cycle. The water in the fibers was even harder to be taken out by the atomization effect than the situation when there was no duty cycle. From the above analysis, it can be concluded that the duty cycle did not affect the water accumulated in the larger pores, but it affected the water accumulated in the small pores. These conclusions can also be arrived from **Figure 10(b)**.

As can be seen from **Figure 10**, it illustrates that when the duty cycle increased from 20 to 40%, the drying time decreased. The drying time also changed when the duty cycle kept the same but the modulation frequency changed. Such as, when the duty cycle was 20%, the drying time was 30 s at 100 Hz modulation frequency, but at 1000 Hz modulation frequency the drying time was 25 s. Therefore, both the duty cycle and modulation frequency of the modulating signal affected the drying time of the fabric.

The effects of different duty cycles and modulating frequencies on the drying time of a 4×4 mm^2 swatch of fabric F-1 on transducer C is presented in **Figure 11a**. The initial water mass was fixed at 10 mg. At a 100% duty cycle, the drying time was 20 s, irrespective of modulating frequency (**Figure 11a**). This value is identical to the drying time obtained when the transducer was continuously driven. The results here demonstrate that the drying time cannot be decreased by changing the duty cycle; however, the power consumed by the transducer can be lowered by doing so [12].

To understand the effects of duty cycle and modulating frequency on fabric drying times, the drying time of a 5 × 5 mm^2 swatch of fabric F-1 was also studied on transducer C. The initial water mass was fixed at 15 mg. The results of these experiments are presented in **Figure 11b**. The results demonstrate that reducing the modulating frequency increased the drying time of fabric [12]. Furthermore, at a given modulating frequency, the drying time decreased as the duty cycle was increased up to a duty cycle D_0. Increasing the duty cycle beyond this value of D_0 did not change the drying time. The results also indicate that the value of D_0 is dependent on the modulating frequency, such that D_0 increased as the modulating frequency decreased. As shown in **Figure 11b**, D_0 = 60% for a modulating frequency of 1000 Hz, and D_0 = 80% for a modulating frequency of 700 Hz. This implies that at a given modulating frequency, the energy consumption can be reduced without affecting the drying time by operating the transducer at a duty cycle D_0.

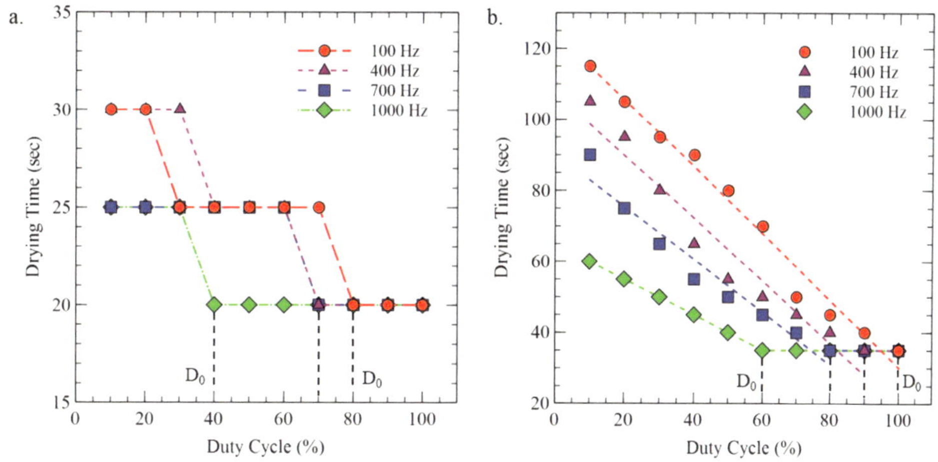

Figure 11.
Effects of duty cycle on the drying time of fabric F-1 on transducer C. Fabric size: (a) 4 × 4 mm^2 and (b) 5 × 5 mm^2 [12].

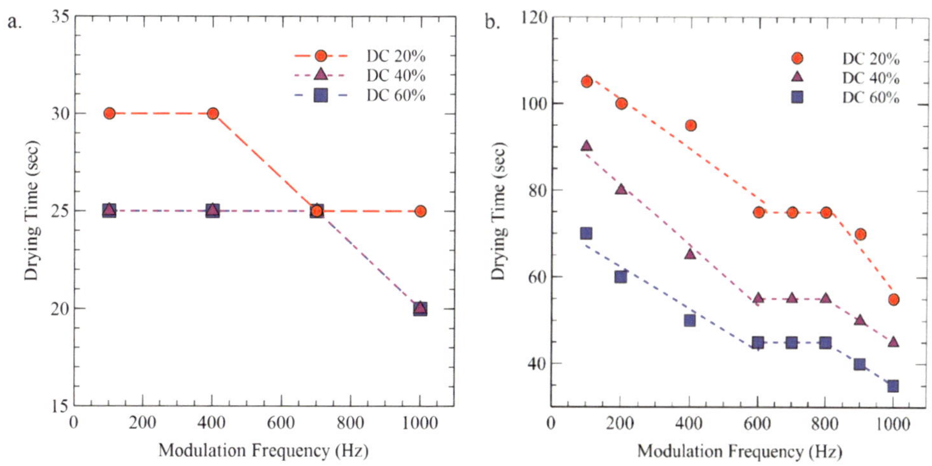

Figure 12.
Effects of modulating frequency on the drying time for fabric F-1 on transducer C. Fabric size: (a) 4 × 4 mm^2 and (b) 5 × 5 mm^2 [12].

The effects of modulating frequency on the drying time of a 4 × 4 and a 5 × 5 mm^2 swatch of fabric F-1 on transducer C are presented in **Figure 12a–b**. The initial water mass for the 4 × 4 and 5 × 5 mm^2 pieces of fabric were 10 and 15 mg, respectively.

As seen in **Figure 12a**, since the difference of drying times at different modulation frequencies were relatively small, such as for the duty cycle 40%, the drying time at 100 Hz modulating frequency was 25 s and at 1000 Hz modulating frequency the drying time was 20 s, the drying time kept the same at certain modulating frequency. But following that modulating frequency, the drying time decreased linearly with the increase of the modulating frequency. Also, **Figure 12b** illustrates that the drying time reduced with the increase of modulating frequency. These conclusions can also be derived from the data in **Figure 11**.

6. Energy efficiency analysis

The EF ($EF_{100-20\%}$) values for the drying of fabric F-1 at different drying conditions were compared (**Table 5**). The duty cycle and modulating frequency corresponding to each drying condition are shown in **Table 5**. Drying type A refers to fabric F-1 drying on transducer A; drying types C1–C5 refer to fabric F-1 drying

Drying types	Duty cycle (%)	Modulating frequency (Hz)	$EF_{100-20\%}$ (lb/kWh)
A	Continuous driving mode		0.27
C1	Continuous driving mode		2.79
C2	10	100	19.01
C3	10	400	19.01
C4	10	700	26.61
C5	10	1000	44.35

Table 5.
Comparison of energy factors (EF) for drying fabric F-1 at different drying conditions [12].

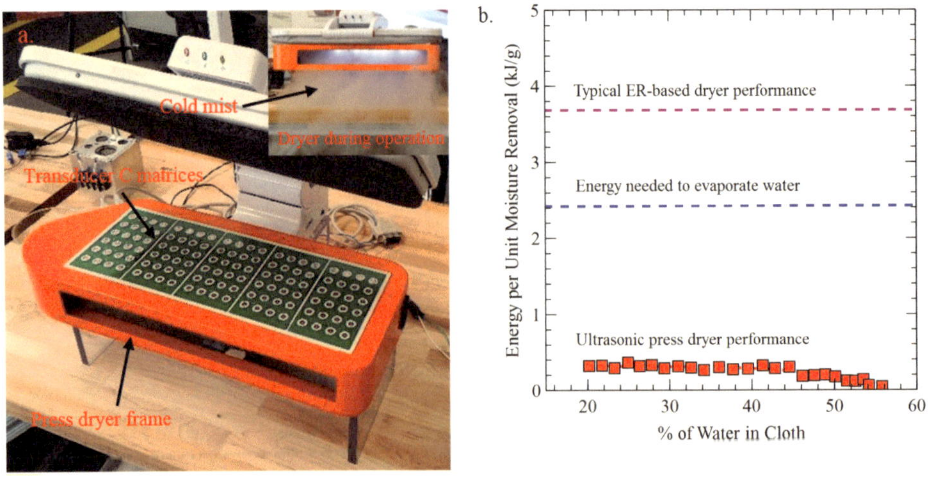

Figure 13.
(a) A mid-scale press type ultrasonic clothes dryer, and (b) its energy consumption data at different initial water content [11].

on transducer C. As illustrated in **Figures 11** and **12**, the drying time for fabric F-1 dropped with an increase in the modulating frequency and duty cycle up to D_0. The $EF_{100-20\%}$ values at different modulating frequencies with the duty cycle fixed at 10% (shortest duty cycle in the tests) were calculated and compared with continuous transducer driving (**Table 5**). The values in **Table 5** show that $EF_{100-20\%}$ values for the new drying method with modulating signals are about 7–16 times greater than those for continuous transducer driving.

A mid-scale press type ultrasonic cloth dryer was designed and fabricated, for which the ultrasonic transducer C was utilized due to its best drying performance among all the three transducers. **Figure 13a** shows the demonstration unit of this ultrasonic press dryer, which a shirt size cloth can be dried by this unit. The energy consumption per unit mass of water content (kJ/g) removed from the fabric by this ultrasonic press dryer was tested. The energy efficiency of it was compared with the latent heat of evaporation for water (at ambient temperature and pressure) and the typical electric resistance (ER)-based dryer (**Figure 13b**). The results show that the energy efficiency of the ultrasonic drying process is much higher than the thermal evaporation and ER-based drying method. Compared with the typical ER-based dryer, the energy efficiency of this ultrasonic press dryer is more than 10 times higher.

7. Conclusion

In this chapter, high-frequency, direct-contact vibrational drying of fabrics using ultrasonic transducers have been introduced as a promising technology to reduce both the drying time and energy consumption of the drying process. The physical mechanism of fabric drying was comprehensively analyzed. The results showed that the fabric drying processes consist of nonlinear and linear regimes, which were dominated by ultrasonic vibration and thermal evaporation, respectively. The electric power could be supplied to ultrasonic transducer in burst mode rather than continuously, saving energy but not affecting the drying processes. Operating the transducer on a shorter duty cycle and at a higher modulating frequency was more energy-efficient for fabric drying. Based on the initial tests on a press type ultrasonic clothes dryer, this direct-contact ultrasonic drying technique showed energy efficiency more than 10 times higher than a typical electrical resistance dryer, and 5 times higher than the latent heat of evaporation at water content greater than 20%.

Acknowledgements

The author would like to express sincere gratitude to Prof. Saeed Moghaddam in Department of Mechanical and Aerospace Engineering, University of Florida, for his support during the research work in University of Florida.

The author also would like to acknowledge support from Oak Ridge National Laboratory for its assistance.

Conflict of interest

The author declares that there are no affiliations with or involvement in any organization or entity with any financial interest or non-financial interest in the subject matter or material discussed in this manuscript.

Author details

Chang Peng* and Saeed Moghaddam
Department of Mechanical and Aerospace Engineering, University of Florida, Gainesville, FL, USA

*Address all correspondence to: chang.peng@ufl.edu

References

[1] How Much Electricity Do Appliances Use? Off Energy Effic Renew Energy. 2015. Available at: https://energy.gov/energysaver/appliances-and-electronics

[2] Denkenberger D, Calwell C, Beck N, Trimboli B. Analysis of potential energy savings from heat pump clothes dryers in North America. CLASP. 2013:1-42. Available at: http://clasp.ngo/Resources/Resources/PublicationLibrary/2013/Clothes-Dryer-Heat-Pump-Technology-Offers-Substantial-Cost-and-Energy-Savings-for-North-America

[3] U.S. Environmental Protection Agency. Market & industry scoping report: Residential clothes dryers. Energy Star. 2011:1-18. Available at: https://www.energystar.gov/sites/default/files/asset/document/ENERGY_STAR_Scoping_Report_Residential_Clothes_Dryers.pdf

[4] Zegers FTS, Molenbroek EC. Field Test of Heat-Fed Washing Machines and Tumble Dryers. Utrecht: ECOFYS; 2000

[5] Lekov A, Franco V, Meyers S. Evaluation of energy efficiency standards for residential clothes dryers in the USA. Energy Efficiency. 2014;7:133-149

[6] Caldeira KG, Chan AK, Hyde RA, Kare JT, Myhrvold NP, Whitmer C, et al. Energy efficient dryer systems. US 9091015 B2; 2015

[7] Lakkineni VR, Zentner MM, Anikhindi SM, Froelicher SB, Jr ON, MTB, et al. Energy efficient cycle for clothes dryer. US 8991068 B2; 2015

[8] Martin E, Sutherland K, Parker D. Measured performance of heat pump clothes dryers. ACEEE Summer Study Energy Efficiency in Buildings. 2016;**2016**:1-13

[9] Yoon C-H. Microwave clothes dryer. U.S. Patent No. 1988;4:765,066

[10] McCarthy DE. Apparatus for drying clothes or other solids using microwave energy under reduced pressure with energy recovery while avoiding arcing. U.S. Patent Application No. 2013;13(/783):198

[11] Peng C, Ravi S, Patel VK, Momen AM, Moghaddam S. Physics of direct-contact ultrasonic cloth drying process. Energy. 2017;**125**:498-508

[12] Peng C, Momen AM, Moghaddam S. An energy-efficient method for direct-contact ultrasonic cloth drying. Energy. 2017;**138**:133-138

[13] Deepu P, Peng C, Moghaddam S. Dynamics of ultrasonic atomization of droplets. Experimental Thermal and Fluid Science. 2018;**92**:243-247

[14] Peng C, Moghaddam S. Energy efficient piezoelectrically actuated transducer for direct-contact ultrasonic drying of fabrics. Drying Technology. 2019:1-10

[15] Patel VK, Reed FK, Kisner R, Peng C, Moghaddam S, Momen AM. Novel experimental study of fabric drying using direct-contact ultrasonic vibration. Journal of Thermal Science and Engineering Applications. 2019;**11**:021008

[16] Jaffe B, Cook WR, Jaffe H. Piezoelectric Ceramics. New York: Academic Press; 1971

[17] Khmelev VN, Savin II, Abramenko DS, Tsyganok SN, Barsukov RV, Lebedev AN.Research the acoustic cloth drying process in mock-up of drum-type washing machine. International Workshops and Tutorials on Electron Devices and Materials EDM—Proceedings. 2006:223-228